CW01558638

PROFIT FR

A GUIDE TO WORKING WITH CARS
and MONEY SAVING TIPS

ALL MOTORING SERVICES

PROFIT FROM CARS

First published 1995 by All Motoring Services

Copyright 1995 All Motoring Services. All rights reserved. No part of this publication may be reproduced or transmitted in any form or by any means, electronic or mechanical, including photocopying, recording, or any information storage and retrieval system, without express permission in writing from the publisher.

ISBN 0 9524339 8 2

Printed by Bell and Bain Ltd., Glasgow

British Library Cataloguing in Publication Data. A catalogue record for this book is available from the British Library

All Motoring Services,
KA13 7QZ.

CONTENTS

WORKING WITH CARS:

APPENDIX:

INTRODUCTION

For some, cars are a hobby; even an obsession. In an ideal world, we would all like to earn a living doing what we most enjoy, and in my case, I am lucky that I do!

This book is written for those who would like to start a career working with cars, those wishing to dabble, or those who just have an interest. Subjects covered are diverse - from dealing in registration numbers, to making a profit from children's pedal cars.

Part One is a complete guide on buying or selling a car with confidence. Use this section for private transactions, or in the course of your business. It is suggested that Part One is read in its entirety, to gain maximum background knowledge for **Part Two**, which relates to the motor trade.

The section **'Working With Cars'** suggests possible career opportunities. Based on personal experience, and that of family and friends, ideas listed could be tried out initially, on a part time basis, whilst remaining in (relatively) safe employment.

Useful addresses and money saving hints are contained in the **'Appendix'**, which also includes helpful tips on running a business.

PROFIT FROM CARS also features a comprehensive handbook on used car dealing, the truth about how some of the more shady operators work, and even includes suggestions for earning extra 'beer money'.

Spray painting, body repairs, and major mechanical work is omitted. These may seem obvious options, but they involve time served skills - apprenticeships requiring years of hands on experience. To encroach on these areas would encourage 'cowboy' operations, which in the long term benefit nobody. Skills mentioned in the book (eg. driving instruction or engine tuning) also take time to master, but the relevant chapters explain what is involved, and how to get started.

Aimed primarily at the man in the street, PROFIT FROM CARS is not a 'money for nothing' gimmick', nor is it a 'get rich quick' book. It is assumed the reader is not starting out with a large financial backing. Hopefully, the text explains to the novice in a clear manner, without being insulting to the more experienced.

References to addresses and telephone numbers are given numerically (bold type in brackets) - match the numbers to those listed in the address section on page 146.

Anyone can be the 'boss' - you don't have to work for someone else; the final chapter in this book offers constructive advice about going self employed, which hopefully, should set you off on that all important road ... *to success!*

PART ONE - BUYING AND SELLING A CAR

BUYING A CAR - GENERAL
Buying a car has been described as the biggest purchase you will make (excepting a house). Mistakes can cost a lot of money, or even result in a serious road traffic accident.

Get clear in your mind what your requirements are. List the following points if buying new:-

a) *Decide what you can afford, consider buying price and running costs - how will it be financed?*
b) *List required specifications - make and model, engine size, colour, 3 or 5 door, manual or automatic transmission?*
c) *Different dealers make different deals! Seek out those in your area.*
d) *Find out what price extra accessories are; does the cost include fitting?*

If considering used:-

a) *Decide on, and stick to your price range.*
b) *As above, list suitable models.*
c) *As well as running costs, check out spares availability and prices. Note that insurance premiums may be out of proportion to the car purchase price.*
d) *Find out how much MoT and tax remains.*
e) *How will you buy - dealer or privately? If you intend buying through auction, refer to the second hand car dealing section later in this book.*

DIESEL or PETROL?
Diesel engined cars offer more miles per gallon at a cheaper price per litre than petrol. They require less maintenance, and should last longer. There is less chance of breakdown due to electrical faults, and depreciation is generally less. Despite vast improvements over

recent years. these engines tend to be smellier, noisier, and need more specialised attention when repairs are needed.

DEPRECIATION

When buying either new or used, a prime consideration must be the value of the car when it comes time to sell. Price guides list values at the time of purchase (up to 10 years ago), and their worth to-day. By using this information, you can calculate the rate of depreciation. Accordingly, you should be able to project the amount a current (similar) model might lose in the future.

Generally, depreciation is higher on: foreign cars, especially East European; cars over four years of age fitted with automatic transmission; large engined, or high insurance grouped cars; models recently updated or replaced; high mileage, neglected types - taxis, etc. ; and cars with unattractive colours - browns, beiges, etc.

Depreciation is usually lower on: so called 'British' cars - Ford, Rover, etc. ; cars with manual gearboxes; currently produced models; economical, and low insurance grouped cars; diesels; light mileage, or one owner, well maintained cars; and cars with popular colours - red, black, white, silver, or metallics.

Note that although a car depreciates rapidly during the first years of its life, by the time it is around nine years old, depreciation will be slight - if any at all.

NEW CARS

Although it is very nice to own a new car, you will find the economics are discouraging when buying for private use. If a new car costs £8,000 and is sold a year later at auction for £5,000, then you will need to think about it very hard! Some people buy new in the belief that they will get a reliable car. In fact there is probably as much chance of this car being returned to the dealer for 'teething troubles' or routine adjustments, as the likelihood of trouble from the year old model. The latter should be rid of these problems by the time the second owner takes possession, and the chance of mechanical failure due to wear or fatigue is highly unlikely. Add to this the fact that the sharpest depreciation has occurred, then the second hand car becomes an attractive proposition.

When considering a new buy, the 'ex-demonstrator' may be of interest. It is basically a new, but low mileage car, which will possibly be registered under the garage name, and used by potential customers for test drives. The big advantage is the reduced selling price, while retaining most of a new car's benefits. Note also that old stocks of models which have recently been updated or replaced generally sell cheaper.

Bear in mind that you should get a better deal at particular times of the year, eg. late June or July - just prior to the issuing of the new year letter in August. December is also a good time to buy, things are very quiet and dealers are keen to make a sale.

Expect, and always ask for a discount - especially easy to get if paying cash (ie. with no finance). Find out if your discount is being deducted from the **basic** price, or the **total** price. If it is deducted from the total (which includes additions such as delivery, number plates, road tax etc.), then your discount will be better than that taken off the basic price (which renders the same percentage less effective). Get the salesman to write down his calculations, and ask for a copy. Make sure that you are not being overcharged for any of these added extras.

Avoid, if you can, trading in your old car. A main dealer may give you a (falsely) high offer for it, but this is only eating into the discount you would have been due anyway with no trade in. Usually he sends older trade ins to auction where they are sold at low trade prices.

'Phone or visit other dealers and brokers who can supply the particular model you are interested in. Get a price, then quote it to the opposition - they will probably try to better it. Play them against each other for maximum benefit!

CAR BROKERS
Office based, rather than at the showroom, the car broker has few overheads. He will not want to take your old car as a trade in, but will shop around to get the keenest price for your new car.
Make sure any car got through him meets U.K. specifications. If you have any doubts about using his services, get a price from him, and see if a main dealer will match it.

IMPORTING FOR PERSONAL USE
Cars bought overseas may be cheap, but if you intend to import one for your own personal use, some careful calculations will be necessary. Any benefit gained might be offset with the cost of trips abroad, red tape, and trying to locate a suitable model which is up to U.K. specifications. Note that some foreign dealers apply a surcharge on R.H.D. models; also your V5 registration document will betray the vehicle's origin (which might not be beneficial when the time comes for you to resell).

Check that any manufacturer's warranty applies to the U.K. If bought in the European Community it should be effective in any member state. Politics play their part; your local franchise dealer might not like what you have done. You may find that if your imported car needs repairing under warranty, there just happens to be "a couple of months waiting list at the moment..."

The Irish Republic is a good source. A member of the E.C., cars are R.H.D., and up to British specifications. There is no language barrier, and main dealers are familiar with personal exports. They will help, or even organise everything for you. Good percentages may be saved on the more expensive models. Remember to take account of foreign exchange rates when working out your calculations.

Some importers get stuck trying to unravel Value Added Tax complications; you might wish to contact a V.A.T. registered U.K. dealer who would import on your behalf (using a letter of authority from you), for a fee. Thus the exporter makes a profit, the U.K. dealer gets his commission, and you make your savings without V.A.T. problems.

Use your local library (or international directory enquiries) to contact main franchised dealers abroad. Note that the major motor manufacturers are aware that this import trickle may turn into a flood, so rules may change to make personal imports impractical. Check the current situation with the exporting dealer.

EXPORTING FOR PERSONAL USE
It is possible to export your new car tax free provided it is for personal use. For further details contact your main franchise dealer. If exporting a new Ford contact the address given at the end of this book [1].

NEW CAR WARRANTIES
Ask the salesman to explain fully the terms of the warranty, not forgetting the 'small print'. These terms vary with different manufacturers - some even offer a change of vehicle if you are not entirely satisfied within a specified time limit. There may be set conditions. You may be bound to have your car serviced at specific intervals, and all repairs carried out by the franchised dealer. Don't be tempted to save money by going elsewhere, as this may render any guarantee void. Excluded from the warranty will be faults caused by general wear and tear.

For a fee, an optional extended warranty might be offered. It is worth considering, but find out which parts are *not* covered, and if it is labour inclusive. Anti corrosion warranties will cover 'rot', but may exclude external surface rust such as stone chipping.

A warranty might be extended by the dealer if the vehicle has been off the road for a lengthy time due to repairs carried out under guarantee. Note that you have no automatic right to a substitute replacement while your car is being repaired, that will depend on the garage. If any right were to exist (under certain circumstances) this would depend on the law relating to damages.

Any remaining manufacturer's warranty should be transferable to subsequent owner(s) of the car.

FINANCING YOUR BUY

Make sure you are buying something you can afford! **The best discount is obtained if you are paying 'up front' without any finance,** even if it means borrowing elsewhere to do this - you will be in a stronger bargaining position. If you do need finance, you don't have to accept the dealer's offer of credit; a better deal may be arranged through your bank, building society, or a finance company.

If choosing a credit deal tied to the car (ie. hire purchase), as distinct from a general loan used to buy anything, you have the advantage of being able to seek assistance from the credit company in the event of having problems with the car which the dealer will not, or can not sort out. A disadvantage is that you are not free to sell the car (if it is still subject to the agreement) without making arrangements with the H.P. company.

You might be offered 0% finance by the dealer, ie. monthly payments but with no interest charges. Usually these deals are offered over a shorter repayment period, combined with a higher initial deposit. Make sure you accept on the *price negotiated with the salesman,* as opposed to the full recommended selling price.

Try to avoid putting up your home as security when borrowing. Check any agreement is covered by a 'cooling off' period - in which you will have a set time to change your mind about entering into the contract. Seek professional advice before making a commitment.

THE A.P.R.

Make a point of finding out the Annual Percentage Rate (A.P.R.) charged on any loan. Generally, the lower it is, the better the deal. It should be inclusive of extras, such as setting up charges, etc. You have a right to ask for an A.P.R. quotation when applying. Before going ahead, make sure you know:-

a) *The A.P.R., the monthly total payable, and how long the agreement lasts.*
b) *The total cost of the credit.*
c) *If you can pay off the loan early without penalty - avoid loans where you still pay the same interest total, even though you repaid early.*

d) The consequences of missed payments due to sickness, redundancy, etc. (some loans offer optional protection schemes).

Your choices if not paying outright may be one or more of the following:-

BORROWING FROM FAMILY OR FRIENDS
BANK PERSONAL LOAN OR OVERDRAUGHT
LOAN FROM A FINANCE HOUSE
BORROWING AGAINST A LIFE POLICY
USING A MONEYLENDER
HIRE PURCHASE AGREEMENTS
CAR MANUFACTURER'S FUTURE VALUE DEALS
LEASE AGREEMENTS

BORROWING FROM FAMILY OR FRIENDS
A cheap option if you are lucky! Beware though, it may seem a great idea at the time, but many a friendship has been lost this way.

BANK PERSONAL LOAN OR OVERDRAUGHT
Contrary to rumour, the bank manager *is* human! Banks are reasonably flexible in that you can usually pay off the loan early without penalty, or even be extended over a longer period. They will look for some sort of security before granting the loan, and may also charge a setting up fee.

Personal loans are usually for larger sums lent over a fixed period. Payments are made on a regular basis at a fixed rate of interest. Alternatively the loan might be paid back with a one off payment after an agreed time. The A.P.R. may be less than that of Hire Purchase.

An overdraught is basically overspending on your bank account. Make sure you make prior arrangement with your bank before 'going into the red'. The loan is quite flexible. Charged interest stops immediately on the amount repaid, and is payable only on the remaining balance.

LOAN FROM A FINANCE HOUSE
Finance houses are reputable money lenders. They usually offer a personal loan which can be used for any purpose, not necessarily just

for buying a car. If you wish to pay your loan off early, you may incur a financial penalty - check the contract's small print. The rates may be higher than the bank, but cheaper than hire purchase, although in this competitive age compare all options - you may be surprised.

BORROWING AGAINST A LIFE POLICY

If you have a life policy, it may be possible to raise a loan (up to the 'surrender' value) from the assurance company. An advantage is the possibility of lower interest rates. The company will consider their loan secure, and a bigger advantage is that the loan may be left outstanding until the policy matures, when it will be deducted from the sum payable. Alternatively you may pay back only the interest, and the capital will be deducted upon maturity.

USING A MONEY LENDER

To be used as a last resort - they may offer instant cash, with little or no security required. They can be cripplingly expensive. Good advice would be to avoid them, and save a lot of heartache.

HIRE PURCHASE AGREEMENTS

Most bona fide dealers will offer you hire purchase finance. An initial deposit is normally required, followed by regular monthly payments over a set period. Legal ownership of the vehicle will not pass to you until the final payment, which may or may not be a one off lump sum, has been received.

Traditionally, hire purchase is relatively easy to obtain, but has not been the cheapest method of finance. As mentioned before, present day competition dictates that you shop around for the best deal. A popular offer is the 0% finance package. To make this a viable proposition for the dealer, you will probably pay a higher than usual initial deposit, with regular payments over a shorter period (possibly twelve months), and may also require a substantial final payment.

Hire purchase is usually offered by the dealer at the point of sale. The salesman may receive a commission from the finance company for each deal he clinches. On the strength of this, he may be willing to offer you a better discount on your car.

Your credit worthiness will be checked, and another disadvantage may be that if you pay the balance off early, you could still be liable for all the interest that would have been due over the full agreement period.

CAR MANUFACTURER'S FUTURE VALUE DEALS
The future value deal (or balloon payment) is a recent development with most of the major car manufacturers. When buying a new car, the dealer will work out (and **guarantee**) a conservative projection of the car's value at the end of the agreement period.

Example: A car retails at £12,000. You pay an initial deposit of £3,000, leaving a balance of £9,000. The dealer estimates the car to be worth £6,000 in three years time (at the end of the agreement term). This projected figure will be your final lump sum payment. Your monthly payments are based on the difference between the cost price less deposit, and the final lump payment. Thus you are only paying back £3,000 (which makes for low payments), but you will pay interest on the full balance of £9,000. In effect you are buying on hire purchase, but deferring a hefty final instalment (in this example, £6,000).

To deal with this last payment, the salesman will offer some alternatives:-

a) *Trade in for a new model. If the actual trade in value of your car happens to be more than the projected figure, the monetary difference can be put towards the deposit of your new car.*
b) *Sell the car. Any profit over and above the guaranteed value may be kept by you.*
c) *Return the car to the dealer. Should the car be worth less than the projected value, the loss will be borne by him. However, there may be clauses imposed regarding the mileage and condition of the car you are returning. Usually the car will be worth more than the guaranteed value; the estimate errs on the side of caution.*
d) *Keep the car and pay the final lump sum.*

My opinion is that although the payments are affordable, you are only postponing that final 'day of reckoning'. Let's deal with each choice in turn:

a) *Trade in for new; you are being subtly persuaded to buy a new car possibly sooner and more frequently than you might wish. Your payments never cease.*
b) *Sell your car, keep the profit, if any; this could be your best option if you can realise a good price.*
c) *Return the car to the dealer as the final payment; unless market prices drop drastically, don't do it! If you do, you will have paid a lot of money to have in effect, just hired your car over the agreement period.*
d) *Keep the car and pay the final lump sum; might be a good choice if you can find the capital, and can bear the thought of handing over a lot of money for what is now an old second hand car.*

There may be a nominal setting up fee for the package. Some agents offer an additional maintenance scheme, whereby for an additional premium, expenses for servicing and normal wear and tear are covered.

Above is only a guide of how the system might work, and figures given are only for illustration purposes; variations exist - contact your local dealers for their particular plan.

LEASE AGREEMENTS
Leasing is basically hiring a car (regular fixed payments) over an agreed period. Benefits are a minimum capital outlay combined with the ability to terminate the agreement after a specified time. Title of the car does not pass to you, but you might be given the chance to buy the car at a favourable price at the end of the agreement period.

If you happen to be a business user, your monthly payments will be tax deductible (obviously as you have no title to the car there is no writing down or depreciation allowance against tax).

Balloon lease agreements are monthly payments made less due to the fact that you will make a large final payment at the end of the leasing term (see 'future value deals' above).

Contract hire is a more expensive lease, but the charges provide for 'free' servicing, repairs, and sometimes road fund licence or R.A.C./A.A. cover. **Check small print,** and be wary of mileage clauses.

USED CARS

DEALERS

The biggest advantage when buying from a dealer, apart from wider selection, easy finance, and service backup, is the fact that they are governed by the Trade Descriptions Act, breaching of which is an offence.

Main dealers sell cars from one or more manufacturer. Their reputation has to be maintained, and as such any used cars bought from this source will be late, low mileage models. It follows that these vehicles will be more expensive to buy, but will be subject to a better backup service than that of cars bought elsewhere. Older, high mileage cars that are traded in will probably be disposed of by the main dealer via the auction house.

Used car dealers (garage forecourts etc.) might retail cars with more variation of age and mileage, and are generally less expensive than similar cars offered by the main dealer, but the service backup up may not be as good as that offered by the main dealer. The exception to these statements could be the dealer who specialises in one particular make, or type (ie. vintage, sports, etc.).

At the bottom of the league is the used car 'lot' or even worse the private person who 'sells the occasional car as a hobby'. The former will probably deal in older, more dubious vehicles, service backup may be non existent - he may be very reputable, but be careful. The latter type of dealer should just be avoided.

When choosing any dealer, go by known reputation. Make sure he is a member of a trade association such as the R.M.I. (Retail Motor Industry Federation) or S.M.T.A. (Scottish Motor Trade Association). Members operate under a strict code of practice. Look for A.A. or R.A.C. inspected cars offered for sale. Note that these inspections are generally less detailed than those carried out by these organisations on behalf of private prospective purchasers.

PART EXCHANGES

Do I or don't I? Trading in is an easy option, but generally if you can avoid the part exchange deal, do so. You are more likely to realise a better price by selling privately, but this may take more time.

As advised earlier, try not to pay the full asking price for any car! The dealer is prepared for this, and so with this in mind, his asking price will be artificially high. If you offer a car as part exchange, it will be set against this higher figure. An extreme example of this would be; you buy a car selling at £4,000. If you paid cash you might get a discount of £400. On the other hand you offer your old banger, worth £300 as a trade in. The dealer gives you £400, and you are delighted in the knowledge that you have done well - but in effect you have given away your car for free!

The principle illustrated above, although exaggerated, applies universally. Remember that the dealer has to make a profit - after all he is in business. He may consider your car is not up to his usual sale standard, and dispose of it through the trade, often getting a much lower price than you might think.

There is nothing gained in obtaining a fantastic trade in figure for your car, if the one you are buying is artificially expensive. When offering a good quality trade in vehicle, you will generally get a better deal if you approach a dealer who specialises in that particular make or type.

In any case, before starting any negotiations, find out the **real** trade value of your car.

USED CAR WARRANTIES
Make sure the salesman explains fully the terms of any used car warranty. Whilst your legal rights remain consistent, the small print and conditions vary widely. Used car warranties are often a type of insurance policy operated by a company which is independent of the garage where you bought your car.

Find out how long your guarantee lasts, and what is covered. Some warranties will cover the full cost of parts and labour throughout the guarantee period. Some may cover parts only, or parts and labour for a specified time, after which only parts are covered. Find out which parts are **not** subject to the warranty.

Some second hand cars may still carry manufacturer's warranty which could be automatically transferable. These are obviously better buys.

PROBLEMS WITH THE PURCHASE
Having exhausted the obvious channels unsuccessfully, you may have to resort to contacting your local Citizen's Advice Bureau, Consumer Advice Centre or Trading Standards Department (addresses in the 'phone book).

Members of the A.A. and R.A.C. may wish to use these organisations for advice or assistance.

You could contact any trade associations of which your dealer is a member (see the address section at the back of this book). They operate under a code of practice, and in addition offer an arbitration scheme for difficult cases.

If all else fails you will have to take legal advice, and if necessary take the matter to court.

BUYING PRIVATELY

Riskier than buying from a reputable dealer, but less risky than buying from auction, buying privately reflects a middle price when compared to the other two choices.

When buying privately you will have fewer legal rights, so it is important that you keep the original advertisement from which you saw details of the car. Try to get the seller to describe the car **in writing** - this could be vital to your case if you have serious trouble with the vehicle later.

There is no onus on the seller to tell you everything about the car. Ask specific questions, eg."has the car ever been involved in an accident?" Should the seller give an untruthful answer, he may contravene misrepresentation laws - but this would be a civil, not a criminal matter. Take a friend with you as a witness, be warned - ask these questions. Silence does not constitute misrepresentation. A private seller is not subject to the Trades Description Act.

Some dealers try to cheat you by purporting to be private sellers. They might advertise in your local shop window or newspaper. Any dealer is obliged by law to declare the fact that he is trading. Be especially careful if:-

a)   *You are asked to ring a mobile telephone number, some examples being exchanges 0836, 0850, 0831.*
b)   *The same telephone number crops up in other advertisements.*
c)   *The name and address listed on the V5 registration document as the registered keeper is not the same as that of the seller.*
d)   *The seller offers to meet you somewhere to discuss the deal.*

An age old ploy of checking the credibility of your 'private' seller is to telephone in answer to his advertisement, and ask him if the car is still for sale. If the reply is,"which one?", this does not look good!

Checking out the documentation, and how to conduct your test drive is described later in this section. Do not gloss over small faults - when added together they could end up costing you a lot of money.

Find out the selling price of comparable cars being sold in your area (bear in mind condition, mileage, etc.) You can do this by scouring newspaper or car trader type magazine advertisements. Don't skimp on the homework!

A car is sold for a reason. Look for tell tale signs of neglect. It may have been, or expected to be a lot of trouble. Possibly the owner couldn't afford to run it anymore, or the person you are buying the car from is really a part time dealer! However, things may not be so gloomy; the car is possibly replaced on a regular basis. The owner's circumstances may have changed, so he needs a bigger, or smaller car, or maybe the vehicle is being replaced by a company car.

Even the type of person that you are buying from may offer some clues: A farmer or self employed tradesman's car may have been overloaded, kept in a filthy state, abused, or generally neglected. The professional person might have covered a lot of miles, but his car will probably have been well maintained, with no expense spared. The elderly may keep a low mileage, main dealer serviced car. The young male has probably 'thrashed' his vehicle, and unnecessarily 'mucked around' with the engine, while the young female might be totally ignorant of the fact that maintenance is required!

CLOSING THE DEAL

When negotiating a car's purchase, half the battle is won by a strong psychological approach. We are all guilty of the thought, "I'll never find one as good as that again". The truth is there are millions more! *Stick to your price.*

Declare from the outset that you still have other cars to view. This way, you will have time to gather your thoughts after your inspection without making a commitment. If you are desperate to buy, you can make an acceptance call from the 'phone box around the corner!

Be pleasant, try to appear calm and level headed. If the seller senses that you are keen to buy, he will have the psychological advantage, and may try to hold out for the full advertised price.

Your bargaining position is improved by pointing out **genuine** faults you have found, but don't be pernickety, or list non existent problems with the car. This will only make you lose his respect, and make his resolve stronger.

Always pitch your bid lower than the asking price. You have nothing to lose - the vendor will probably expect this, and has no doubt priced the vehicle accordingly. An example being a car advertised for £2,150. The seller is obviously expecting a 'lazy' bid of either £2,100 or £2,000.
In this situation, try bidding less than the expected figure, ie. £1,900 or £1950. If your offer is accepted, you have made a *genuine* gain.

Avoid giving the 'take it or leave it' ultimatum. He may call your bluff, you can't re-negotiate without loss of face, and in the meantime someone else has made an acceptable offer. A better idea is to offer him a realistic price, should he be reluctant to accept, tell him you will leave the offer open for a specified period after which you will contact him again. If he hasn't had a better offer by that time, the chances are that he will be more willing to talk business!

PRIVATE SELLERS AND PURCHASERS WARRANTY
A contract scheme which can protect both the private buyer and / or seller is offered by Nationwide Used Car Arbitration Ltd. (N.U.C.A.).

A simple pro-forma contract is provided which is legally binding. Should a problem still be unresolved 21 days following the sale, arbitration will be provided at no further expense to the contract holder. The scheme is backed by A.A. Technical Services. For further details of the scheme, contact N.U.C.A. [2].

Information about the private buyer's rights is detailed in the relevant section in the chapter SELLING A CAR.

CHECKING THE POTENTIAL PURCHASE

DOCUMENTATION
The car may be sold with or without a **road tax** disc. Check the expiry date, and that the details, ie. make, model and registration number all match the vehicle.

The **insurance document** relates to the owner or driver of the car, and is thus non transferable.

THE REGISTRATION DOCUMENT V5
Insist on seeing the vehicle **registration document**, also known as the V5. Check all details are correct, matching the Vehicle Identification Number (V.I.N.), engine number, colour, registration number, and the recorded mileage (if listed) at the last change of vehicle keeper.

If the registered keeper's name is different from that of the seller, find out why. In the case of a discrepancy, seek a reasonable explanation. Take note of the amount of previous owners, and also check how long the seller has owned the car. Some self drive hire companies and well known driving schools may be registered on the V5 with a different, less well known name.

Some people may advertise their vehicle as a '1994 model'. They may be bending the rules a bit - the car may have 1994 specifications, but actually first registered in 1993. Be sure to note the date and country of first registration.

To sum up; a V5 document should be available for inspection, and come with the car. I would tend to be suspicious if this were not the case. An omission can be an easy way of hiding a car's history, or might indicate a recent change of keeper. At worst is the possibility of a stolen car!

THE M.O.T. CERTIFICATE
Note the expiry date, that it is valid for the correct vehicle. and the certificate is endorsed with the testing station's embossed stamp.

An M.O.T. is required for any car that is three years old or more, and must be renewed annually. The test, conducted only by approved garages will only assess the condition of environmental or safety components, and therefore does not pertain to the general condition or reliability of the car. The certificate only relates to the condition of the car **at the time of the test.**

The recorded mileage is logged on the M.O.T. slip at the time of testing. This may help verify the true mileage, even better if the owner can produce old certificates and / or any documented service history.

If considering a car that is without a certificate, note that you may need to spend lot of money to bring it up to the required standard of the test. However, these M.O.T. failures are generally cheaper to buy, and may be of interest to those who are capable of carrying out the necessary repairs.

Ask the seller if he still has the test failure slip. This will list the defects, giving a rough idea of the amount of work needed and the expense involved.

When offered a car that has less than six months M.O.T., it is worth asking the seller if he would be willing to put the car through another test at your expense. For a relatively small fee you will have extra 'peace of mind'. Speak to the tester and tell him that you are interested in the car, and ask for his comments at the end of the inspection. In the event of the car failing, you might wish to make a revised (lower) offer for the car.

CHECKING THE MILEAGE
Average annual mileage on most family cars is around 10 - 12,000 (more for diesels). High mileage cars could be considered provided they have been well maintained and the asking price is in accordance with it s condition.

'Clocking', the practice of falsifying the odometer (mileometer) reading, is still one of the most widespread frauds in the motor industry. A late model, high mileage car can be bought, clocked, and sold for half as much again - very lucrative for the dishonest seller.

Compare your car with another of similar age and mileage, and look for clues. Some examples may be worn pedal rubbers, a shiny polished steering wheel, sagging driver's seat (compare with the passenger's side). Make a note of any wear and tear uncharacteristic to a car of the supposed mileage, ie. oil leaks, radiator appearance, or stone chipping on front body panels. Check especially the area around the speedometer. Retaining screws may be burred, odometer figures may not be displayed in an even line, which indicates that the instrument has been tampered with.

It can be a hard job to detect clocked cars which still show a high mileage, ie. was 120,000, now showing 78,000, as these vehicles will be showing some of the wear symptoms described above.

Ask the seller for the registration document. and note the name and address of the previous owner. Contact him via directory enquiries, he may help you regarding mileage, previous history, etc. Try to inspect the service booklet if one exists. Dealer stamped entries will verify mileages, the level of maintenance received, when, and how often.

HIRE PURCHASE INFORMATION
If it came to light that you had bought a vehicle in good faith, not knowing that it was the subject of a hire purchase agreement, you should still get title to it. However, you have no protection if the car is subject to any lease or hire contract. You could check that your car is not on hire purchase (or been an insurance write off), by contacting your Citizen's Advice Bureau, or a company called Hire Purchase Information Plc (H.P.I).

For private car buyers, H.P.I. will check if a car is reported as being stolen, subject of a substantial insurance claim for accident damage, had a previous cherished number transfer, or is subject to an outstanding hire purchase agreement. A one off fee is payable for this service.

For further details, contact H.P.I. via their private purchaser's telephone number [3].

THE CAR INSPECTION

GENERAL

Make sure you do your homework. Get hold of old test drive reports from car magazines, and find out the strengths and weaknesses of the model you are interested in. The *'WHICH? Guide To New And Used Cars'* is available from the Consumers' Association [4], or your local library. It offers independent assessments of reliability, depreciation, etc.

If you are unsure about checking your intended purchase, seek outside help, such as a knowledgeable friend or mechanic. You could use your local garage, the R.A.C. Vehicle Examination Service [5], or A.A. Vehicle Inspections [6].

Note: Any vehicle inspection should be carried out within a reasonable space of time. In the list which follows, comprehensive directions are given for inspecting most components. It would be impractical to carry out all the checks listed within the time available. However, if you have suspicions about any particular aspect of the car, detailed instructions are given for a further in depth check.

Make allowances for normal wear and tear - remember you are not buying a new car. You may wish to make a more detailed inspection of the car (at home), immediately after the purchase.

CHECKLIST

For reasonable efficiency, arm yourself with the following items:-

>   The seller's name, 'phone number, and address
>   Change for the telephone (or take a Phonecard)
>   A copy of the advertisement, and a price guide
>   Pen and paper (for defect list and notes)
>   Torch, magnet, jack, axle stands, towrope, petrol can
>   Old bedsheet, cardboard, or similar to lie on
>   Overalls or old clothing, and cleaning rags
>   Inspection list *as follows ...*

BODYWORK

The structural condition of the vehicle bodywork is vitally important. An engine is relatively easy to replace, but in the event of severe accident damage or advanced corrosion, repairs may not be a viable proposition.

Inspect the car in daylight - rain can hide poor paint finishes and evidence of repair. View the car from a distance, is the car sitting level? Any listing indicates worn suspension, distortion, or severe corrosion. All panels such as doors, boot, front wings, etc., should sit square to the adjoining bodywork, any gaps being parallel. Next, from the front of the car, look down both sides in turn for dents or scratches. Ripples, or light scratching, points to previous body filler repairs. Use your magnet to confirm, you know it will only adhere to metal and not body filler, glass fibre etc. which may lurk under the paintwork. The owner may have placed a sticker over a scratch or botched repair.

Rusted bodywork and accident damage can be hidden by a respray. Look for clues - overspray on wheels, tyres, trim, and glass. Prise up the rubber trim which surrounds the front and rear screens, or if this is not possible, open a door and lift a little part of the inner trim. Check these areas for old paint which is a different colour (do take into account any natural fading). All exterior panels should be the same shade of colour.

Open the car bonnet, inspect the paint, but also look for one or more newly fitted components ie. headlights, radiator, bright metal fittings / screws, or any other indications of a front end bump.

Some traders make a sad car look like new by the simple use of a quick respray, often with little or no preparation work. Around six months after the car has been sold, the bodywork tends to revert to its former state!

Genuine resprays which have been properly prepared, may have been done to brighten up faded paintwork, for a change of colour, or to repair minor damage. A little known fact is that nearly all cars are subject to some degree of accidental damage during their working life.

UNDERBODY
Find out the most vulnerable areas prone to corrosion on the make and model that you are going to inspect.

If you can obtain the use of a pit for the inspection, all the better. Failing this, raise the car securely with a **hydraulic trolley jack**. The car should rest on **axle stands**. If you must use portable ramps, ensure the car is secure by leaving it in gear, handbrake applied (if possible), and the car is stabilised with **wheel chocks**. The use of an inspection lamp is ideal, second best is a good torch. Avoid smoking when working under the car.

Check the underbody for rot, and use the *handle end* of a large screwdriver to probe suspect areas. Look for previous repair work, paying particular attention to sills, inner sills, jacking points, and all suspension mountings. Inspect the petrol tank and fuel lines for leakages. Check the exhaust system for corrosion, leaks, temporary repairs, and loose or missing mounting brackets.

The existence of old, original underseal is encouraging, while any that has been recently applied could be covering either a skilful repair, or (horror) body filler, or glassfibre botch ups.

Newly fitted suspension or body components (especially if all are fitted in the one area) should be investigated. Search for other evidence of an accident, eg. panel distortion or heavy denting.

Upon the completion of the underbody inspection, open the car boot. Lift the carpet to check for dampness or rusting. Note the condition of the spare wheel. Open the car doors, and see that the doorskins are securely attached to the frames, especially at the lower half, where rainwater may have gathered. Open the bonnet to inspect the paintwork, strut mountings, and the inner wings.

Steering, suspension, and brake checks are described later.

GLASS
Inspect all glass for cracks, chips, or windscreen wiper damage. M.O.T. regulations have been extended to cover windscreen condition.

Most are now bonded in place, making replacement beyond the scope of the average enthusiast. With rising costs, it may be worthwhile enquiring if repair, rather than replacement, is possible. Read the small print of your insurance policy, windscreen damage may be covered.

INTERIOR
A fresh, well cared for interior is a sign that the car has been looked after. Alternatively, years of misuse or neglect will be hard to hide, and might indicate that the rest of the car has suffered the same fate.

Test the seat belts (front and rear)for efficient operation, fraying, and security of mountings. The sunroof should be watertight and work properly.

Look for yellowing of the headlining caused by smokers, and inspect the seats for burns. Open the boot or tailgate for evidence of heavy usage by tradesmen - scratches, cement dust, sawdust, hay or straw, etc.

Water ingress can be exceptionally hard to locate. Watch for dampness or rotting of the carpets, possibly caused by worn seals, badly fitting panels, or advanced corrosion. When the source has been found, the carpet can be replaced by cheap and readily available non original items.

SECURITY
Make sure all locks and catches operate smoothly and correctly. Steering locks and central locking systems can be expensive to repair or replace. If the car has been fitted with a burglar alarm, this is a bonus, check for efficiency.

The chassis (V.I.N.) plate is usually to be found within the engine compartment. Ensure the number is correct, and the plate has not been obviously refitted or tampered with.

Any etched registration number on the glass should tally correctly.

ELECTRICS

All lights, exterior and interior should be checked, as should the operation of switches, gauges, horn, heated rear screen, wipers, heater, and all accessories.

The starter motor should turn the engine over briskly. Do the same again, but with the headlights on. If the starter motor is sluggish, there is a bad earth, internal motor fault, or the battery is below par. If the battery is suspect remove any oxidisation that may have built up on the terminals (using a wire brush or hot water).

Any roughness indicates wear on the starter pinion or flywheel ring.

Ensure that the battery (ignition) warning light is extinguished upon starting the engine. Even when idling there should be no trace of any glowing, which would suggest an alternator fault.

SUSPENSION

The vehicle should appear to sit level when viewed from any angle. If this is not the case, inspect for broken or worn springs.

Shock absorbers should be free of fluid leaks and severe corrosion. Wear is indicated if there is a jarring of the car when negotiating rough roads or potholes. If they are excessively worn, the car might suffer constant vibration, or more commonly, wallow or sway when being driven. A simple efficiency test can be carried out by pressing down firmly on each corner of the car in turn. When released, the body should move up, down again, and then stop. Further movement should not continue. Listen carefully whilst carrying out the test; knocking or groaning noises might mean worn mounting bushes, or other related problems.

Road springs or shock absorbers should always be replaced in pairs, not singly.

Other suspension faults are indicated by unusual noises, wandering, or pulling to one side when driving or braking, and heavy steering (make sure tyre pressures are correct).

STEERING

'Free play', ie. how much the steering wheel is turned before the roadwheels start to move, should be within the tolerances set by the car manufacturer, up to about 3-4 centimetres should be acceptable in most cases. Excess movement signifies a worn steering rack or joints. Operation should be almost silent, without any grating or roughness.

An assistant under the vehicle while you are carrying out this test may see or hear any cause of the free play. He could confirm by grasping the suspected component, feeling for slackness or knocking as the steering wheel is being rocked back and forth.

Check that the two rubber gaiters (one at either end of the rack) are not cracked or leaking oil.

If the car wanders, or pulls to one side when being driven, this could also indicate suspension or steering faults, such as incorrect wheel alignment. Steering wheel 'shimmy', or vibration, points toward suspension / steering wear, or roadwheels out of balance. See the next section regarding alignment and wheel balancing.

Cars that are equipped with power steering should be checked as above. The steering wheel should be turned full 'lock' both left and right (ie. to the end of its full travel). Feel for any snatching or vibration, movement should be progressive, smooth, and quiet. Inspect the fluid reservoir, and that drivebelts are taut and in good condition.

TYRES, WHEELS, AND ALIGNMENT

The legal requirement for cars is not less than 1.6mm tread depth throughout the central three quarters of the breadth of the tyre, and around the entire outer circumference. There should be no bulges or splits.

Uneven tyre wear is a clue to possible steering and suspension faults. Wear around the centre of the tyre; over inflation. Worn tread around the outer 'shoulders'; under inflation. If around only one shoulder; poor alignment or other fault. Patchy, uneven wear could be due to worn or damaged suspension, a faulty wheel bearing, brakes, or wheels out of balance. Other associated causes may be worn transmission joints, or a twisted, distorted tyre - commonly occurring with remoulds.

Check the roadwheels for buckling or dents.

Tracking adjustment, or front wheel alignment is a relatively small job, but necessitates the use of specialist equipment, and is therefore best carried out by your local 'fast fit' tyre centre.

Should the car vibrate in a particular speed range, but is cured by driving above or below it, there is a strong possibility that the roadwheels need to be balanced. When experienced through the steering wheel, the problem lies with the front wheels. If felt through the floor or seat, then the rear wheels need balancing. Again, use your local tyre depot, but shop around for the keenest prices to align or balance.

WHEEL BEARINGS
These may be checked by jacking the car up, and detecting any free play by rocking the roadwheel back and forth (one hand at the 12 o'clock position, the other at 6 o'clock). Note that some car manufacturers recommend some slight play. If in doubt, contact the service department of your main dealer. Faulty wheel bearings can also be noticed when the car is being driven, by grating noises from the roadwheel, confirmed by the wheel feeling warm to the touch.

BRAKES
Remove front roadwheels to inspect the brake discs. They should appear bright silver, indicating proper contact with the pads. Some light scoring is acceptable, but heavy indentations will result in scrapping the disc. Brake pads should have at least 2.5mm of friction lining remaining. With most cars it should be possible to view the pads without the need for any further dismantling.

The fluid reservoir should be maintained to the correct level. Inspect all brake pipes and hoses for corrosion, leakages and kinks. Check for any chafing or fouling with the car body.

Press the brake pedal firmly. It should feel solid - any sponginess will probably be due to air in the system. If pumping the brakes eradicates the softness, this will confirm. Investigations will have to be carried out to find the cause. After rectification, the brakes will require bleeding.

Excessive pedal travel is indicative of brake wear. If the pedal starts to sink slowly after holding well for a few seconds, this could signify defective seals inside the brake master cylinder.

Should the brake pedal need excessive pressure to produce results, the servo unit may be at fault. Test for correct operation by first pumping the brake pedal, and then depressing it firmly (engine off). Switch on the engine. The pedal should now sink slightly and then feel firm again.

The brakes should be responsive without pulling to one side (possibly caused by a seized calliper) when the car is driven. There should be no grating noises (worn pads), or juddering (brake, steering, or suspension fault). If the judder is felt through the brake pedal, suspect a distorted disc. Squealing brakes must be investigated, but may be caused just by a build up of dust.

Any brake problems must be thoroughly and immediately investigated. Any of the above symptoms could be caused by reasons other than those listed.

HANDBRAKE
Properly adjusted, the handbrake should hold the car firmly on the second or third click of the ratchet, operating quietly and smoothly.

When driving the car at around 5 m.p.h., apply the handbrake firmly. The car should slow down considerably. Failure to do so could be caused by:-

a) *Poor adjustment*
b) *Stretched handbrake cable*
c) *Cable or linkage needing lubrication*
d) *Brake linings worn or contaminated by brake fluid*
e) *Faulty or seized automatic adjusters*
f) *Handbrake mechanism faulty (inside brake drum)*

CLUTCH

The clutch is often worn out due to misuse or ignorance. Biting point, (when the clutch starts turning the wheels) should be approximately 3/4 way up on the pedal's travel. Most modern cars have an automatic adjuster to compensate for wear.

Faults include: Stiff pedal operation, crunching or difficulty in selecting gear, shudder or fierceness when taking up the drive, and screeching or rumbling noises.

To check for a worn clutch, apply the handbrake firmly with the car stationary, pointing up a steep hill. With the engine racing, try to drive the car up the hill (with the handbrake still on). Hopefully the engine will stall, or even try to drive 'through' the handbrake. Should this not be the case, and the engine still works with the clutch pedal fully up, you have a severely worn clutch!

Inspect the clutch cable for fraying (might cause clutch shudder), and smooth operation. If hydraulically operated, check hose condition and fluid level.

With the exception of the cable or hydraulic mechanism, access to the clutch normally entails a lot of work. Therefore if any of the three main clutch components (pressure plate, friction disc, and thrust bearing) are in need of replacement, it is a false economy not to renew them all at the same time, given the relatively low parts cost.

TRANSMISSION - GEARBOX

Inspect the casing for oil leaks. All gears including reverse, should be easily selected, with no crunching or noises. Bear in mind, symptoms could be caused by a faulty clutch.

When road testing, drive in the low gears for longer than normal, building up the engine 'revs' before changing up. Next, at lower speeds lift on and off the accelerator. These checks should ascertain if the car is prone to jumping out of gear.

The option of a second hand, or reconditioned gearbox is sometimes a cheaper option than getting your own one repaired; professional attention to the gearbox can be expensive.

TRANSMISSION - AUTOMATIC GEARBOX

Make sure when driving under all loads and conditions, that the gears change both up and down when they should, smoothly and quietly. Use all modes; Park, Reverse, Neutral, Drive, 2nd. Gear, and 1st. Gear. There should be no hesitancy in gear selection. A long delay indicates wear. When in Drive mode, at around 40 m.p.h., press the accelerator to the floor. This should cause the gearbox to 'kickdown', ie. the gearbox immediately selects a low gear for instant acceleration.

Test the inhibitor switch by ensuring that it is only possible to start the car when in Neutral or Park mode.

Inspect the outer casing for leaks, then check the gearbox transmission fluid level by using the dipstick. The car may have to be up to operating temperature, with the engine idling - check with your main dealer if this is the case with your car.

The colour of the fluid can give clues of gearbox condition. It should be red in colour. Burnt or dark fluid may mean burnt clutch linings. Sticky, thicker fluid implies overheating has taken place, and the presence of any metal particles could indicate serious internal wear or failure. Note that only specialised attention would confirm any of the above possibilities.

TRANSMISSION - DIFFERENTIAL

Inspect the casing for oil loss. When road testing, listen for any whining (may be acceptable on some models), or rumbling noises (unacceptable).

TRANSMISSION - PROPSHAFT, U.J.'s AND C.V. JOINTS

At low driving speeds, apply heavy on / off pressure to the accelerator pedal. Any slackness in the transmission joints should be apparent by 'clunking' noises.

With rear wheel drive cars, pay attention to the propshaft. If the universal joints are suspected, support the car securely on axle stands, and try twisting the shaft by hand or lever, to detect wear.

Front wheel drive cars are tested for constant velocity (C.V.) joint wear, by listening for heavy knocking when turning sharp corners. There may also be a cracking noise when taking up the drive in first gear, after having previously been in reverse gear (or vice versa).

THE ENGINE

Inspect the engine for oil leaks, or rust stains - indicative of water loss. Look at the external condition of the radiator and hoses. Check the water level, noting if anti-freeze is present. If, on first opening the bonnet, you notice that the engine is warm, you will know if the seller has started up the car before you arrived - not suspicious in itself, but this might camouflage nasty start up noises, or other problems associated with a worn engine started from cold.

When starting up the *cold* engine, listen carefully for dry bearing noises. A cloud of blue exhaust smoke signifies worn valve stem oil seals. Ensure the oil pressure warning light extinguishes almost immediately. If the car is fitted with an oil pressure gauge, you will usually find a healthy reading when the engine is cold, but after the engine has reached operating temperature (and the car is being driven at a steady speed), the gauge should *still* show a constant reading at the recommended pressure. A drop in pressure when the engine idles is acceptable. Low, or fluctuating pressure could be caused by a faulty oil pump, or engine wear, the latter involving expensive repairs.

To check for engine 'back pressure', caused by worn piston rings or cylinders, remove the oil filler cap and increase the engine speed. Evidence would be excess fumes or oil escaping from the filler.

Generally, more noise means more trouble; top end 'tappet' noise may mean at best poor maintenance, and at worst, excessive valve gear wear.

Knocking noises from the engine (usually when under load) are usually associated with worn 'big end', or main bearings - a major job.

Metallic 'pinking' noises, evident when the engine is under load, ie. accelerating, or being driven uphill, are caused by;-

a) *wrong grade of fuel used*
b) *incorrect ignition timing*
c) *engine needs decarbonised*
d) *engine is overheating*

Check that the engine doesn't run on after the ignition is switched off. The causes of this are similar to those of engine 'pinking', as listed above.

There should be no undue noises from any engine ancillaries (alternator, water pump, etc.) If you are suspicious of any particular component, hold it by hand - this may make the noise or vibration more evident, but **be careful!**

Ensure there are no noticeable smells inside the car, such as exhaust fumes, hot oil, steam etc., when the engine is running.

Rough running or idling may mean an ignition or carburation fault.

When driving off, check in the mirror for exhaust smoke. Emissions can tell a lot about the state of the engine, ie. :-

Blue smoke - cylinder, or piston wear; use a compression tester on the engine to verify. If the smoke is evident on the 'over run', this could be due to worn valve stem oil seals.

Black or grey smoke - a rich petrol mixture; check carburettor settings, the 'choke' mechanism, and condition of the air filter.

Steam - If in excess, it is probably caused by a leaking or blown cylinder head gasket, sometimes combined with engine overheating. Make sure the engine has reached operating temperature; a cold engine will produce steam anyway. If the oil on the dipstick is grey, or covered in condensation, this is also suspect. Extra checks are made by looking for oil contamination in the radiator water, or air bubbles appearing as the engine is running. Another symptom is the presence of a creamy 'mayonnaise' substance inside the oil filler cap, or inside the rocker cover.

Before carrying out these checks, note that extreme caution is necessary when removing the radiator cap. At operating temperature, the hot or boiling water is pressurised, and liable to spout out.

The spark plug condition can also provide clues; remove them each in turn, and look at the electrodes:-

A tan or mid grey colouring is the correct colour - good news!

Pale grey, or white - the petrol mixture may be weak, or the engine might be overheating (the latter combined with any blistering of the porcelain indicates severe roasting).

Oily, black, or carbonised electrodes - possibly due to general engine wear.

Wet electrodes (smelling of petrol) - could be too rich a petrol mixture, the plug not firing due to being faulty, or another ignition problem.

ROAD TESTING
Make sure you have adequate insurance cover. The drive should last about three miles, taking in all conditions over different types of road (ie. twisty, fast, and rough). This should bring to light the more prominent faults. You will assess performance, listen for odd noises, and note instrument dials or warning lights.

When the drive is finished, listen for any overheating noises, ie. 'grumblings', coming from the radiator. Leave the car standing, and after a few minutes check for any oil or water leaks.

DON'T GET CONNED!

The common practise of car traders posing as private sellers was referred to earlier on page 17. Unfortunately, this is rife, especially with part timers intending to make a little extra cash, 'on the side'. Most bona fide traders are honest, but as in any group of people, there are those who spoil things. There now follows notes on some of the more common 'cons'.

Some unscrupulous dealers literally join together two different accident damaged cars, in order to make one that is seemingly genuine. These hybrids are known within the trade as 'cut and shut' jobs. The work may be expertly carried out, but cut and shuts are potentially dangerous, and not worth buying. If your suspicions are aroused, contact H.P.I. (see page 22), they may be able to help. When inspecting the car, run your hand along the interior headlining, you might be able to feel a welded join. Does the car look as if it has recently been resprayed? As an extra check, you could *discreetly* scrape off a minute amount of paint from both ends of the car to ensure they are the same colour!

Another problem associated with the car trade is the 'ringer'. A dishonest seller steals an expensive car. He then buys a similar, cheap, accident damaged model, complete with documentation. The chassis number from the write off is then transferred to the stolen car, which he then offers for sale to the unsuspecting public. As previously mentioned, look for new rivets fitted to the chassis plate, or any other signs of faking.

Even if you bought your car in good faith, but it was later discovered to be a ringer, it may still be taken from you, with little possibility of compensation.

Falsifying vehicle mileage is common, refer to the relevant section on page 21. A car's mechanical life is based on mileage, rather than age.

Wiring the oil pressure light to the ignition / battery warning light (or vice versa) has been known, although this practise is less common with the advent of circuit board wiring. This 'con' makes both lights extinguish upon starting up (as they should), despite the fact that the car may need a new alternator or engine!

Deliberately blocked or disconnected engine breather pipes and hoses may stop tell tale oil or fumes escaping, which would normally betray severe back pressure - characteristic of heavy wear and high mileage.

Engine knocking and bearing rattles can be temporarily camouflaged by the use of proprietary thick oil additives, available from car accessory shops.

Always be wary of new looking resprays, many are done for honest reasons, others cover botched repairs. Look under the wings and other vulnerable areas for jagged metal, chicken wire, or even old newspapers used as makeshift bodywork!

The subject of checking M.O.T. certificates was dealt with on page 20. Don't take this document at face value; **it has been known for unscrupulous owners of corrosion ridden vehicles to have their cars M.O.T. tested every 6 months or so. Why? Because if the vehicle failed due to terminal rot, it could be sold on to an unsuspecting buyer, having 6 months remaining on the *original* M.O.T.**

SELLING A CAR

PREPARATION AND PRESENTATION
It is generally accepted that the overall appearance of any car is the major influence in effecting a sale. Good presentation is therefore essential. If a car can even be made *just to look well cared for*, you have won half the battle. There have been instances where a grubby looking car bought at auction has been well valeted, re-entered, and sold at a good profit!

BODYWORK
After giving the car a thorough wash, use a good chrome cleaner for brightwork. Bumpers and trims can be cared for with suitable products such as rubber and vinyl cleaners.

Remove any stickers (eg. "We've been to Wales!"), with caution. Make sure you don't tear any old underlying paint as you do so. Check that you are not going to be left with a faded patch where the sticker was.

Reds and maroons are most prone to weathering. Colour restorer such as 'T-Cut' helps to revive faded paint, but some of these applications contain abrasives, not recommended for cars with a metallic finish. Rubbing compound is even more fierce, use only for severe cases of fading. Applying this with too much enthusiasm will result in a loss of paintwork!

New arrivals on the market are non abrasive restorers, such as Car Plan 'Colour Wax', and Turtle Wax 'Color Magic'. These have an inbuilt colouring agent, so buy a compatible shade to match that of the car. You may wish to finish off with a good wax polish.

Should you intend repainting rusty areas, think carefully first. The tiny rust mark may end up as an unsightly repair; it is not easy to keep the work localised. If the affected area is noticed by a prospective buyer, he may think there has been a cover up, and wonder what else you are trying to hide. Although you know you that you have painted over (just) a slight scratch, he may suspect a more serious problem.

New paint is easily detected, and it may not be possible to achieve a good colour match, especially with metallic paints. When effecting a small repair, use the 'touch in' (with brush) variety, rather than aerosols which tend to blemish a wide area. Should you decide to use aerosols, which are better for larger areas, practice your technique beforehand to achieve a good result.

GLASS
All glass should be well cleaned. Use a non smear window cleaning agent on the inside of the glass. This area is often overlooked when preparing a vehicle, but can improve the appearance drastically, even on the oldest banger! Don't forget to polish the mirror lenses.

INTERIOR
Vacuum clean the interior. Empty and wash out the ashtrays, and if you are a smoker, install an air freshener to rid the car of any staleness. Should the headlining have yellow tobacco staining, consult a main dealer for the best method of cleaning your particular model.

Proprietary upholstery cleaners are very effective for seats and trim. The fascia and dashboard can be made to look new by applying specialised silicone based cleaners, available from car accessory shops.

ENGINE COMPARTMENT
Engine cleaners such as 'Gunk' are good, but a more thorough job is done by having the engine steam cleaned, hardly any dirt is missed. If you are unsure who offers this inexpensive service locally, any garage should point you in the right direction.

A well maintained car inspires confidence with any potential buyer. Check, and top up all fluid levels; oil, water, window washer, brake, clutch, and auto gearbox / power steering if appropriate. Most modern batteries are maintenance free, but the older type may need topping up with distilled water. In any case, a nice touch is to smear the terminals with vaseline (prevents oxidisation). Check the fanbelt for tension and fraying.

Oil or grease all hinges, ie. doors, boot, and bonnet - make sure the bonnet pull works easily.

Ensure all switches work, and any faulty light bulbs are renewed as necessary. Check tyre pressures and condition (don't forget the spare). Test wiper efficiency, and replace blades if worn.

THE BUYER'S RIGHTS (SELLING PRIVATELY)
Contrary to popular opinion, the buyer *does have rights* when purchasing a car privately.

Be sure that you describe the car that you are selling in an honest manner - you don't have to inform the buyer of any existing faults, but you must answer specific queries honestly.

According to the Road Traffic Act, 1988, Section 75, "No person shall supply a motor vehicle or trailer in an unroadworthy condition". This means that if you sell an unsafe car, **even unknowingly**, you may end up in trouble with the authorities. This does not apply if the car is being sold for export, or if you have cause to believe that the buyer would not use the vehicle on the road until it has been repaired (in the case of faulty lights, that you believe the vehicle will not be used at night until fixed).

Copies of the Road Traffic Act may be purchased from Her Majesty's Stationary Office (H.M.S.O.) [7].

See page 19, regarding private sale contracts.

ADVERTISING THE CAR

The Business Advertisement Disclosure Act, 1977 requires all advertisers who sell goods in the course of business (even if selling on someone else's behalf) to make the fact clear in their advert. The word 'trade', or even the letter 'T' may suffice in some cases. Failure to declare so may mean prosecution. If you are not sure, get in touch with your local Trading Standards Office. Alternatively, seek guidance from the publishers with whom you intend to advertise.

WHERE TO ADVERTISE?

Shop Window adverts are not generally very effective, due to low reader circulation, so don't put your card in the window if you are in a hurry to sell. Shop ads are better suited for selling older, cheaper cars. Advertising is inexpensive, and you will be able to include a colour photograph at no extra charge.

Your Car Window could carry your 'for sale' sign or poster - and it's free! Although there is a small readership, the biggest advantage is gained by the fact that any potential buyer can see at first hand the merchandise on offer!

Television advertising has become more accessible to the man in the street, with the arrival of TV text. Despite the national coverage, it probably isn't the most effective medium. Check the advertising rates carefully.

Computer Agencies usually advertise nationally. Their advertisement might not carry much detail about the car, but you might only be charged a one off fee which covers all costs, until the car is sold.

Most agencies are reputable, but be on your guard against unsolicited 'hustlers'. After reading your newspaper advertisement, they might telephone you with details of their 'service'. The agency may give the impression that they will not charge any fees unless the car is successfully sold. Make a point of asking them for written details of their terms of business. Be very careful what you say over the telephone - you could be sent a bill, and held legally responsible by a (possibly) recorded conversation, which you can hardly remember making! Ascertain if any fee payable is subject to a sale, a percentage commission, or if you will be just charged an entry / advertising fee.

The press is generally accepted as the best, and most successful means of advertising a car:

Local newspapers are excellent for selling cheaper cars, especially as buyers will be less inclined to spend time and money travelling outwith their area to view. Advertising fees are relatively cheap, but this medium is generally less suited to selling specialist, or more expensive models. These periodicals are usually published towards the end of the week - ideal for week-end shoppers! Local papers are usually left lying about the house up to a week after purchase, which is to your advantage.

National newspapers are quite good for selling expensive, rare, and specialised cars, because potential purchasers are more willing to travel a distance to get the car that they want. The biggest advantage is the wide readership, although it can cost! One snag with the daily paper is that it is generally thrown out by the reader the day after publication.

In my opinion, the *auto market type magazines* are the best bet for your advertisement, when selling privately. The majority of its readers are interested in car buying, or are enthusiasts (so the possibility exists for an impulse buy!). You will have a very wide circulation of readers, most living within reasonable travelling distance. The modest fee usually includes a 'free' photograph entry. You may not be charged per line, but will probably be limited to a maximum amount of words. Various sizes of photograph and advert will be available, dependent on how much you want to spend. An advantage is that the magazine, and therefore your advertisement, may be kept for weeks before being binned.

'Exchange and Mart' provides wide circulation, combined with national coverage, at modest cost. Ideal for advertising expensive, or rarer cars, but not very practical if selling the average family saloon, as buyers will be reluctant to travel for viewing.

General *marketplace type newspapers,* provide great circulation figures, and in your area. These papers sometimes offer free advertising - what have you to lose?

Monthly car magazines (the 'glossies'), offer national readership. Excellent for advertising specialised marques, more so if the magazine is related to the type of car you are selling - you are liable to achieve a higher asking price than would be the norm elsewhere. This type of advertising may be expensive, and is not really suitable for selling cheaper vehicles. The magazine will possibly be kept for years by the reader!

THE ADVERTISEMENT
Make your advert (and thus your car) stand out from the crowd, to beat the competition. It is worth paying extra to have your advert framed by a box, or make the wording catch the eye by using asterisks, or bold type. Don't skimp on the wording, especially if the vehicle you are trying to sell is expensive - generally, the more you are asking for the car, the more you should be prepared to pay for advertising.

Probably the most annoying aspect of selling a car is the problem of so called buyers who are simply time wasters. Always state the selling price; asking for offers will attract a lot of calls which are both wasteful and disappointing. Your geographical area should be included alongside the telephone number, this will sift out buyers who are not willing to travel for viewing. A recent trend is for advertisements to include the statement, "no time wasters". This might seem like a good idea, but can be counter productive. It could intimidate a genuine buyer - he won't know before seeing the car whether or not he will buy it, so he may ignore your advert, and approach a less hostile seller. How do you know that your last 'time waster' wasn't a genuine buyer who just didn't like your car?

You could state,"no agencies" in your advert. Hopefully this will eradicate the advertising companies who build up your hopes by replying to your advertisement, only to have them dashed by a sales spiel for their services.

Avoid using a box number for replies. Most people are impatient, and a telephone enquiry is both immediate and easy. Always have your telephone manned. If enquirers constantly get the engaged tone, they will soon give up. Until customers have actually seen what is on offer, your car will be no better than the many others advertised.

WORDING
Always start off the advertisement with the make and model of your vehicle. Basic information, such as age, price, and a contact telephone number is essential.

Use your descriptions to maximum advantage. If there is a long period of road tax or M.O.T. left to run, say so. Should the remaining period be short, simply state, "taxed", or, "M.O.T.'d". The same would apply to fitted extras, such as stereos. When expensive, inform the advertisement reader - if cheap, ignore the fact!

Use the year of manufacture, or the registration number year letter (issued in August), to advantage. In the case of models first registered between August and December, state *the letter, not the year*, in the advertisement. If registered between January and July, mention *the year of manufacture, not the year letter.*

When describing the car's appearance, choose your words carefully. Phrases such as "no rust", or "as new", are contradictions in terms for a second hand car - all used cars have *some* rust. A dissatisfied buyer could make life difficult for you, and you may be in breach of misrepresentation laws. Safer descriptions, such as "attractive", or "clean car", would be more acceptable.

When a potential buyer reads an advertisement for an elderly vehicle, he will be concerned about possible corrosion. Put some reassurance in your advert, ie. "recently restored", or "solid body" (if appropriate).

Expensive, non standard accessories should be included in the advert, but even if your car happens to be no better than any other similar model, it's image will be enhanced by mentioning so called 'extras', ie. sunroof, central locking, airbag, etc., even though they were fitted as standard equipment by the car manufacturer!

Should the car be a popular colour, such as red, black, white, silver, or gold, etc., include this in your wording. Surprisingly, colour influences the final decision of many potential buyers. At least, if the colour doesn't suit them, you will be spared some fruitless enquiries.

Having spent money on mundane repairs, it is generally a waste of time to mention the fact. Nobody will be interested if the steering has just been fixed - they expect it to be in good working order anyway! In contrast, expensive components recently fitted (such as a new engine, or a stainless steel exhaust) should be included in the advert, as they inspire expectations of long, trouble free service.

Use the time honoured retailer's trick of avoiding round numbers when setting the asking price. Forget £900, make it £895. For some reason, this psychological trick really does work! Remember also, to pitch the price higher than the figure you really want. Your buyer will no doubt make an offer, and generally his bid will round off your asking price. An example being; you want to realise a figure of £2,000 for the car. State your price as £2,295, and nine times out of ten, your customer will round off the figure to £2,000, which he perceives to be a brutally low bid - but he has really paid the figure that you wanted. You could be even luckier with a timid (rounded off) bid of £2,200, and gain an unexpected bonus of £200!

Try to match the mood of your advert to the type of vehicle that you are trying to sell. For example, sports cars suit an urgent, 'buy now' approach, the classic lines being, "not for the faint hearted", or "real driver's car". These phrases really mean nothing, but they captivate certain mental images. Prestige, or executive cars need a cool, relaxed 'civilised' manner, without economising on the description, in order to create an illusion of a beautiful limousine (see below).

EXAMPLE ADVERTISEMENTS
Here are some exaggerated examples of imaginary car advertisements, followed by observations and criticism.

Fiesta 1.1, 'G' registration. M.O.T.'d one year, taxed. Rosso red. Clean car, one lady owner. Stereo. No agencies, £1,895 ono. Telephone 0123 (Barnforth) 456.

Notes: 'Ford' was omitted from the text, everyone knows they make Fiestas. More space is left for other information. The *year letter* was included, as opposed to the *year of registration*. The car was bought new in August, 1989. Had the car been registered in the following January, I would have stated '1990'. It has a long M.O.T., but there is only one week of road tax remaining! Red, being popular may help a sale. The car is clean, but the gearbox is on it s last legs, and the engine rattles. The lady owner is a driving instructor, and she used the car for tuition. The stereo cost £15 in 1977! She doesn't want unnecessary telephone calls, so the price is there, also her home town, and she doesn't want to be contacted by computerised car agencies.

TVR Tasmin 1988 model, 8,000 recorded miles. Shattering performance, must sell this week. £9,950, no offers. Tel. Sometown 987654.

Notes: A hard hitting advert, which suits the car. The car was registered new in 1987, but has 1988 specification. *Recorded* mileage may let the seller off the hook if found out later to be incorrect. "Must sell this week", means buy now, before somebody else does! No lower offers for the car will be accepted, because the car would normally sell around £12,000. The seller pitched his price low, it has instant attraction, and the figure he ends up with is more predictable.

Jaguar XJS convertible, 1994. This one owner car has been meticulously maintained by Jarrats of Park Lane, with no expense spared. Finished in midnight blue, over silver. Expensive sound system, alloys, and central locking. A rare example of a much sought after model, offers - Plentymuny 256

Notes: This laid back advert generates the picture of a lovely car. The wording implies it is not going to be sold for £500, so enquirers will be prepared for a heavy price tag. In this instance, 'offers' suggests discretion.

RESPONDING TO ENQUIRIES

You have done your homework before putting the car up for sale. By consulting price guides, and scouring the 'cars for sale' adverts, you have decided on your asking price. Provided the figure is reasonable, try to stick to it, even if it means taking a little longer to sell. Re-advertising your car at £12 per week over a month is considerably cheaper than reducing your price by £300 - £400, to achieve a quick sale. However, if your vehicle is re-advertised for more than a couple of months, it will become recognised as a car that can't sell, and could put off potential buyers.

Keep notes of all inquirer's telephone numbers, so that in the event of a deal falling through, you will still have a list of potential customers to contact again.

When answering an enquiry, ensure a firm time is set for viewing. This means you won't have a full day wasted waiting at home for a visitor who might not even appear. Let him know that there are others interested in the car - he will then want to be the first to view, and try to beat the opposition.

The buyer may lose confidence in the car if he thinks that he is the only viewer - why is nobody else interested? He might consider himself in a good bargaining position, and submit a low offer. Alternatively, he could become complacent, and delay his visit until the week-end - but by then he may have lost interest, or be tempted elsewhere.

Let him think your car is too good to miss; half the battle is convincing the inquirer to actually come and see the vehicle.

THE POINT OF SALE

Before the viewer arrives, park the car out of sight, ie. behind the house, or in a lock up. By doing this, there is no possibility of your visitor driving up to the house, looking at the car, and going home without even knocking at your door!

When you meet him, be pleasant, and try to appear confident. Give the impression that the car will easily sell. Notice if he seems enthusiastic - should he be really keen on the car, you will be in a stronger bargaining position - possibly to the point of holding out for the full asking price!

The prospective buyer will want to inspect, then test drive the vehicle. Should he wish to take it out on the road, ensure he has adequate insurance cover, and accompany him on any drive.

Do not lie about your car, it is not worth it. In fact, it sometimes pays to mention minor faults (which the purchaser would probably find anyway). He will be more inclined to trust you if you do this, and the more confident he is in you and the car, the more likely it is that he will buy it. Should he inspect the car and subsequently find faults that you hadn't mentioned, he may wonder what else you are keeping from him.

Without exception, do not even consider selling a dangerous, or potentially dangerous vehicle.

Point out any favourable aspects of the car, ie. low mileage (if high mileage, mention *nothing*). Praise the car, but don't overdo it. Should the buyer mention any faults or shortcomings, reply that you took these in to consideration when deciding on the price.

CLOSING THE DEAL
Try to keep a level head when haggling, bear in mind there will always be another buyer for your car. Remember that virtually everyone who buys will expect to make an offer, so don't be offended! Keep in your mind that minimum figure which you will accept. As previously mentioned the asking price is more than the figure you realistically expect to get. The idea is that you achieve this figure, but the customer goes away smiling because he successfully managed to 'beat down' your price.

I have a friend who doesn't even look at car advertisements until they are at least two weeks old; the idea being that he will be in a strong bargaining position for any as yet unsold car, and make a cruelly low bid! With this example in mind, should *your* car remain unsold for any period, you could say that the car had all but been sold earlier, then the deal fell through when the buyer could not raise the cash.

Assuming you have had a successful sale, don't promise to keep the car unless you have been left a substantial, preferably non returnable, deposit. In the event of the deal falling through, the amount should be enough to cover all expenses including repeat adverts. Write out a receipt stating that the deposit is non returnable if the buyer reneges on the deal, and get him to agree by signing it.

Without a deposit, you may not see the buyer again, because he will have time to reconsider, and may have got 'cold feet'. Meanwhile, you will have turned away other enquirers, and when you do re-advertise there is the stigma of having a car that people may think is difficult to sell for some reason.

Do not accept payment by instalments. This is risky, unreliable, and leaves you wide open to legal problems.

Keep both the car and its V5 until you have been paid in full. Holding the V5 by itself as security is not a good idea, because the purchaser can easily apply for a duplicate. Cash as always is king, next best being a building society's **own** cheque, or a banker's draft. When paid by any cheque, hold the car until it has been cleared. You may be able to pay your bank a special fee to speed up the clearance process.

PART TWO - PROFIT FROM CARS

ALL MOTORING SERVICES

PART TWO - PROFIT FROM CARS

USED CAR DEALING - GENERAL

Before reading part two, hopefully you will be familiar with the previous chapters, which dealt with the general aspects of buying and selling a car.

Many books have been written about buying or selling a car, but most of these publications have been aimed at the private individual. Relatively little has been written about the general 'ins and outs' encountered by the motor trader. Relevant information would necessitate the space of a full book, but hopefully this guide will impart enough knowledge for you to make a start in the trade with some reasonable degree of confidence.

There are various methods of trading, dependent on your experience and expertise, or personal circumstances, financial and otherwise. You could begin by attending car auctions and retailing cheaper cars, but there is generally a higher risk of 'come backs' when selling to the public. Alternatively you could become a trade dealer, by purchasing stock from the trade, selling on mainly through auctions. Whilst this may be a lot safer than retailing, it is reliant on high turnover, because each individual transaction carries a low profit margin. Another option might be dealing only in exclusive vehicles, such as exotics, sports, and rare classics, etc. The recommended retail prices for theses cars are vague, so you have more scope for potential profit. This means that the possibility exists to buy (with more confidence) from private sources, and sell via auction, trade or retail.

Permutations of the above methods of trading are, of course, possible.

RETAILING

Selling to the general public is probably the most obvious method of used car dealing, but it is also the option most subject to problems imposed by regulations, consumer's rights, etc. However, when successful, it can be both lucrative and satisfying.

In any type of business, you need to know exactly what you are doing. Mistakes can be expensive. Do your homework; get to know specifications, particular traits, and the price ranges of vehicles with which you intend to deal. If a customer makes an enquiry which you are unable to answer, tell him you will find out, and get back to him as soon as possible. People generally have an inbuilt mistrust of used car dealers. It is up to you to build up an honest reputation, and instil your clientele's confidence. Try not to be unfair or overbearing in your dealings, and aim to be sympathetic, but be necessarily firm when negotiating.

The first hurdle in clinching a deal, is persuading the customer to test drive the vehicle, because, (a) you have established that he has a serious interest in the car, and (b), having taken up your time with the test drive, he won't want to be classed as a time waster - the ball has started rolling, and psychologically he is now keen on the car.

Any vehicle sold should have at least 11 months M.O.T. remaining, and a minimum of 6 months road tax. Don't forget the obligatory stereo (or radio at least).

Selling any car is in theory very simple. Unfortunately when retailing legitimately there are complications and red tape to overcome. Some of these are:-

a) *Trading Standards Authorities*
b) *Local District Councils (Planning and Licensing)*
c) *Customs and Excise (V.A.T.)*
d) *Capital and cashflow availability*

TRADING STANDARDS AUTHORITIES

When selling to anyone, you should be able to do so with a clear conscience. Remember that you are dealing with people's lives, as well as their cars. If you knowingly (or unwittingly) sell a car that is dangerous, you may end up in serious trouble with the authorities. The bona fide retail car trade is geared up in favour of the private buyer; cover yourself by detailing on the sales invoice any known faults with the vehicle.

As a motor trader, you are subject to many rules and regulations, some of which are briefly noted here. For specific details of any law, contact Her Majesty's Stationary Office (H.M.S.O.) [7].

Advertisement Disclosure Order, 1977. When offering goods or services in the course of business (even if on behalf of someone else), you must make the fact clear in your advertising. Failure to do so constitutes an offence.

Consumer Credit Acts apply when vehicles are sold under any finance agreement.

Sale of Goods Act, 1979 / Misrepresentation Act, 1976 / Trade Descriptions Act, 1968. The vendor must have legal entitlement to sell. The goods should be free of any outstanding hire purchase agreements (although exceptions do apply). They must be accurately described, be of merchantable quality, and be fit for the purpose for which they were bought.

Road Traffic Acts, 1972, 1974, 1988. Detailed laws concerned with drivers, and the safety of motor vehicles (refer to the first paragraph on this page). You may also wish to familiarize yourself with prevailing *Construction and Use (C.U.R.) Regulations.*

LOCAL DISTRICT COUNCILS - PLANNING

Before applying to your local council for a dealer's licence, you would be well advised to approach the planning department, with a view to obtaining planning permission for your business. You are less likely to have this granted if you intend trading from a council, or private house - even less so if it is rurally located.

Certain conditions and restrictions may be imposed on how you run your operation, but each application is considered on its own merit.

After submitting your application, you may be required to advertise your intentions to the general public. This may take the form of an advert in your local newspaper, or a notice displayed at or near the intended premises. Objections may thus be lodged by members of the public, for the council's consideration.

Remember that you may be subject to commercial rating tariffs.

When choosing a site for retailing popular family cars, you should look for a prominent place in or near the town centre. Should you wish to retail more specialised vehicles, such as classic cars, competition will be less fierce, and you will be more likely to receive enquiries from further afield. With this in mind, the fact that you are not relying on passing trade, means a town centre site will be less important, and therefore cheaper premises in more obscure locations will be more acceptable.

Enquiries regarding planning permission should be addressed to the Director of Planning at your local council offices.

LOCAL DISTRICT COUNCILS - LICENSING
Before setting up as a second hand car dealer, you will have to apply to your local council for a trader's licence. Terms and conditions for the issuing and continued use varies throughout different areas of the country. The following criteria should be regarded as a very rough guide.

Operating your business without the necessary permission can result in a hefty fine. Approach the council in your area for a copy of any published guidelines on used car trading. These directives should lay out your legal obligations.

The licence fee may vary, dependent on whether you trade with, or without premises; how often it is renewable is also dependent on your council's policy. A notice of intention to trade may have to be displayed (see above).

Business records will obviously have to be kept, but the requirements for V.A.T. purposes, such as *invoices* and a *stock book*, will probably suffice (see below).

You may receive official visits to ensure your operation is being run in accordance with regulations. The issuing of *used vehicle pre sales inspections* may be compulsory.

LOCAL DISTRICT COUNCILS - PRE SALES INSPECTIONS
The licensing authority may insist that a pre sales inspection is carried out before each vehicle is offered for retail sale. The report should be made available to potential customers, the best place to display being on or inside the car. The customer should also be issued with a copy upon purchase of the car.

Requirements may vary, but the vehicle's details, and mileage declaration / disclaimer will have to be included, as will a comprehensive list of tested items. Work carried out, and defects outstanding, should be entered (if none - state, 'none'). The following checklist may be of assistance:-

TYRES - check for wear / condition / appropriate fitment / condition of spare.

BRAKES - shoe or pad wear / efficiency / pipe & hose condition /fluid leaks / handbrake operation and condition.

STEERING - check joints for wear / steering play and leaks / wheel alignment.

SUSPENSION - condition / shock absorbers - check for leaks.

FUEL TANK & LINES / HYDRAULICS / FLUIDS / check for leaks.

COOLING SYSTEM - condition / anti-freeze strength.

EXHAUST - mounting condition / leaks/ corrosion and efficiency.

ELECTRICAL - accessories / switches, etc. / battery condition.

LOCKS - check efficiency and condition /door pulls / window winders / operation of seat adjustment / seatbelt efficiency (front and rear).

BODY - chassis / suspension mountings, check for corrosion or damage / check external appearance and condition.

The car should be road tested, and the following checks carried out:-

Efficient starting / performance / oil pressure / correct operating temperature / engine noise when cold or hot / exhaust emission / clutch operation / gearbox noise, selection and operation / driveshaft and universal joint noise / axle noise / steering operation / suspension and brake operation and efficiency.

Upon completion of the road test, extra checks should be made for oil or water leaks.

The name of the person who carried out the inspection should be entered. Upon completion of a sale, the signatures of both buyer and seller should be dated and logged on the document.

V.A.T. - USED CAR DEALER SCHEME

Before commencing trade, apply to your nearest Customs and Excise (Value Added Tax) office for a copy of their free information booklet, 'Notice 711 - Second Hand Cars'.

Working with V.A.T. can prove complex, so the notes that follow should be treated only as a rough guide - you would be well advised to consult your accountant for assistance and up to date advice.

Should your business turnover exceed (or might exceed) the threshold figure set by Customs and Excise, you must notify your local V.A.T. office, so that you may be registered. If this is the case, you will be issued with a registration number which must be displayed on all your sale invoices. As a motor trader, you will be obliged to issue *tax invoices* for sales, *tax receipts* for purchases, and keep an official *stock record book.*

To explain how the system works, first think of buying goods from a shop. You will be charged the basic price plus V.A.T. at the current rate. If you were charged the same amount of tax on a car, the sum involved would be excessive. Due to this problem, a special tax scheme for second hand car dealers is allowed by Customs and Excise. The V.A.T. charged by the dealer is not on the car's price, but on the **gross profit** (*that is the profit before deductions for repairs and other expenses*). Because the tax due is calculated this way, the trader's price remains competitive. Also, the customer will not be able to work out what profit the dealer is making on the transaction, because V.A.T. charged is not itemised on the sales invoice.

Should a loss be made on the car, there will be no taxable profit, so no V.A.T. will be due, however the dealer can not offset the loss against other profitable deals.

Unfortunately, commercial vehicles, ie. vans, etc., fall outwith the special scheme, so the full rate of V.A.T. must be charged on the basic selling price.

THE V.A.T. PURCHASE RECEIPT
When operating under the special Customs and Excise scheme for used car dealers, you as a trader, must issue any private seller with a purchase invoice (receipt) for each vehicle that you buy for stock.

Use a triplicate invoice / receipt book - the top (original) copy is kept for V.A.T. records, a copy given to the seller, and the remaining part is kept for your own records.

The following details should be recorded on the invoice / receipt:-

Date of the transaction, receipt number, and total price paid

The sellers' name and address, your name, address and V.A.T. registration number (if applicable)

Day book (your records), and V.A.T. stock book number (keep in numerical order)

Car details; date of first U.K. registration, make and model, registration mark, engine and chassis (V.I.N.) numbers

The private seller should certify that he has sold the car at the stated price by signing and dating the receipt. He should also verify the mileage declaration which will be included on the receipt, ie.:

"The mileometer reading is xx,xxx. The mileometer reading is correct / incorrect (delete as necessary).

The approximate true mileage (if different from that listed above) is xx,xxx.

I do not know if the mileometer reading is correct or incorrect (delete as necessary)."

Both parties sign and date the invoice.

THE V.A.T. SALES INVOICE
The same criteria applies to the sales invoice as the purchase receipt. The following details should be recorded:-

Date of the transaction, invoice number, and total price paid (inclusive of V.A.T.)

The buyer's name and address, your name, address, and V.A.T. registration number

Day book (your records), and V.A.T. stock book numbers (keep in numerical order)

Car details; date of first U.K. registration, make and model, registration mark, engine and chassis (V.I.N.) numbers

You should declare, *"Input tax has not been and will not be claimed in respect of the car sold on this invoice"*. This statement should be signed and dated by you, the dealer.

Include your mileage declaration / disclaimer:

"The mileometer reading is xx,xxx. The mileometer reading is correct / incorrect (delete as necessary).

The approximate true mileage (if different from that listed above) is xx,xxx. I do not know if the mileometer reading is correct or incorrect (delete as necessary)".

In addition, if you are accepting a trade-in, or swap, the details of the other vehicle should be given:

The monetary allowance made against the other vehicle.

Date of first U.K. registration, make and model, registration mark, engine and chassis (V.I.N.) numbers.

V.A.T. stock book reference number (also day book, if appropriate).

Both parties then sign and date the invoice, the buyer acknowledging receipt of the car.

THE (V.A.T.) STOCK BOOK
As most used car traders will achieve a high annual turnover, it is assumed that you as a trader, will be V.A.T. registered, and taking advantage of the special tax scheme offered by Customs and Excise. Accurate records of transactions will have to be kept in the form of a stock register. Even if you are not registered for V.A.T., you may still be obliged to keep a stock book to satisfy the terms and conditions of your trader's licence.

Your stock book should record the following details:-

Stock number, in numerical sequence

The date of purchase

Purchase receipt number

The seller's name, address, and V.A.T. registration number (if applicable)

Vehicle details; registration mark, make and model, engine and V.I.N. numbers, date of first registration in the U.K., and the last registered keeper's name and address

The date of sale

Sales invoice number

The buyer's name and address

Any interim change of vehicle details (ie. engine number, etc.)

Purchase price, sale price, and profit margin on the sale

Current V.A.T. rate at the time of sale, and the total sum of V.A.T. due on the transaction

As well as the above requirements for tax purposes, you may also be obliged to list:-

Colour of the vehicle

Date of last M.O.T.

Mileometer reading at the time of purchase

Note that V.A.T. records must be kept for at least six years; in certain circumstances (eg. lack of storage space etc.), this time may be less **if agreed by arrangement** with Customs and Excise.

Standard Forms Ltd., of Romsey [8] can supply you with stock books and pre-printed sales / purchase invoices.

General information on Value Added Tax is given later in this book.

CAPITAL AND CASHFLOW AVAILABILITY

The capital outlay involved when car retailing is daunting. Even a modest display of eight cars costing £3,000 each, represents a total of £24,000 - and remember, no investment is without risk.

How you raise capital is a personal matter, and is dependent on your own circumstances. Retailing cars on a low budget is possible but the chances of success are much enhanced when the venture is well financed.

Bear in mind the usual business running costs, such as heating and telephone bills, mandatory insurances, wages, advertising fees, and trade subscriptions.

Renting, rather than buying your premises will release much needed capital. If you are just setting out, you could find it worthwhile to rent vacant forecourt space from an established petrol filling station. You could negotiate an arrangement with the owner whereby he receives a percentage figure of your sales in lieu of monthly rent, thus improving your 'cashflow'. Some traders take this a stage further, using their site to sell cars on behalf of the customer. These vehicles are entered and displayed for a one off fee, on a sale or return basis. This means that the dealer has virtually no money tied up in stock, and stands to make a profit *whether or not the car is sold.*

Another advantage of retailing cars that you are usually paid 'up front', and so avoid the bad debts usually associated with other sales and service industries. Even when a car has been sold by hire purchase, the H.P. company not only pays the sale price, but also the relevant commission. Although the selling of cars will be the main source of income, there are various sidelines (such as hire purchase, warranties, or even shop sales) which all contribute towards a healthy turnover.

CAR PRICE GUIDES

A wide range of used car price guides are available through newsagents. These books are quite good for the private buyer, but the professional dealer will use confidential trade guides; the well established **"Glass's Guide"**, and the **CAP "Black Book"**. The difference in the prices listed in the genuine trade guides, and those available to the public, can be as much as the profit you would expect to make on one transaction!

Both Glass's Guide [9] and the CAP [10] Black Book are supplied strictly trade only, upon receipt of an annual subscription. You will have to provide proof that you are a bona fide trader, so your application should be made on headed writing paper. Information imparted by the guide is deemed confidential, and is therefore not to be divulged to third parties.

Price guides that are available from your newsagent are much cheaper than their trade counterparts. Some, like the 'Motorists Guide', published by Foxpride [11], are also available by annual subscription.

Whilst all guides list the popular marques, certain exotic or specialist vehicles, such as Ferrari, Lamborghini and TVR, may be omitted. Price guides for classic or collector's cars sometimes appear in the relevant monthly magazines.

CAP also publish a 'Car Identification Book', which provides details of model updates, visual identification features, mechanical specifications, chassis numbers, and V.I.N.s. Also available is their 'Index Of Vehicle Registrations', a reference book detailing dates of registration marks, and their issuing authority. Both books are published annually.

JUDGING CAR PRICES

It is impossible for any price guide to be accurate for every individual car, so adjustments will have to be made for condition, mileage, etc. Some guides offer details on how to take these factors into account.

There can be price fluctuations dependent on geographical area - even differences between rural and city locations. As always, do your homework. Glean local newspapers, auction guides, and even other garage forecourts to get the feel of prices in your part of the country.

Seasons of the year produce market price changes, springtime and summer are traditionally busy, while winter and late autumn remain more depressed.

The difference between December 31st., and January 1st. is one year! The latter date used to be the best time to buy new, as the customer could get twelve months usage before the next new year registrations. To avoid a 'hang or burst' situation occurring at the end of each year, the motor trade now has *two new years*, the first in January, and the second being in August, when the new *year letter* is issued - this is now the most popular month for buying new.

Economic depressions generally create a drop in new car sales, which can actually be of benefit to the used car market. Severe slumps affect all areas of the trade.

New vehicle launches, or 'facelifts', tend to bring down the second hand value of the models they are replacing.

Generally, new cars that command high prices depreciate steeply during the first few years, but the price 'bottoms out' and remains stable after around twelve years.

Supply and demand is the basis for pricing anything. A good example of this being brand new cars (with a customer waiting list), which go to auction, and achieve higher bids than the recommended retail price!

CHOOSING STOCK - FROM WHERE?

Part exchange deals, the motor trade, and auction sales are all prime sources of suitable stock. As these three options are dealt with elsewhere in this book, this section is concerned with buying from the general public (most commonly through advertisements placed in the press).

Should you follow a classified 'car for sale' advert, you are unlikely to be offered it at a price lower than would be attained at an auction. To achieve a good deal you will have to be 'thick skinned', and be lucky enough to be at the right place, at the right time, with the right seller! To beat the price down to an acceptable *(for you)* level will require skill and diplomacy; the seller may feel insulted! However, the time and effort put in can reap handsome dividends.

You may wish to place a general 'cars wanted' advertisement in your local newspaper. This strategy can also produce good results. Any potential seller who answers your advert will likely be (a) short of cash and keen for a quick sale, (b) ignorant of the way car dealers operate, or (c) is not aware of his car's value. Be reasonably fair when making your offer, especially if you suspect any of the above examples apply. Should you be blessed with a sensitive nature, this method might not be best for obtaining stock!

There are advantages gained when buying from the public. You will have a good opportunity to inspect the car thoroughly, have a test drive, and ascertain any history by checking service records, garage receipts, old M.O.T. certificates, etc. You may also be able to refer to the previous keeper before you buy. Find out his details from the V5, and use directory enquiries to establish contact.

CHOOSING STOCK - BEST BUYS

Don't fall in to the trap of buying cars that *you* fancy - remember you are trying to run a business. Any car is a good purchase, provided the price is right, but a bad buy in the early stages of trading can prove disastrous.

When building up your operation, keep to simple, popular family cars. Refer to the article on page 4, regarding depreciation. Listed are the vehicles to avoid, and those worth stocking.

Commercial vehicles, such as vans, are best avoided when buying from trade and auction, unless you are V.A.T. registered, or the seller is not charging V.A.T. on the sale. As these vehicles are not subject to the Customs and Excise dealer's scheme, they will be subject to the full rate of tax.

ACCEPTING TRADE INS

An advantage for the customer in buying from a dealer, is the option of using his old car as part payment. He will be excited about the prospect of his new purchase, but will have little enthusiasm for his old car - for which he may have good reasons for getting rid. After road testing the new vehicle, he will be more aware of his old car's faults, and he knows that easy disposal is readily available. This is one of the few times that a trader can make a low offer for a car, but still keep the customer happy!

If the car you are retailing carries a high enough profit margin, you will be able to afford a generous offer for the trade in - you might even give more than the market value under the right circumstances.

Disposal of these part exchange vehicles might be effected by retailing them from your premises (you might be lucky enough to make a profit out of *both* cars). You may prefer to dispose through the trade for a nominal figure, being happy with the profit made on the overall transaction. All cars are worth money, and if the trade in is not a retail proposition, you could put it to auction.

WARRANTY AGREEMENTS

Most customers buying from a dealer will expect a guarantee on their vehicle, over and above their basic legal rights. The warranty usually covers mechanical failure, but it may have wider scope.

Some dealers may undertake major repairs independently, but most will use a bona fide warranty company, ie. a mechanical insurance scheme. Warranties encourage sales - if your profit margins are sufficiently high, they could be price inclusive. Alternatively, the guarantee package could be sold as an optional extra.

Varying types of warranty cover may be available, different types of policies could apply to new cars, vehicles up to five years old, seven years old, and cars of any age. Some of these guarantees may be transferable upon a later change of owner - an attractive selling point.

If you wish more information, the addresses of some warranty companies appear at the end of the book [12].

HIRE PURCHASE AGREEMENTS
A high percentage of car buyers deal with retailers primarily because it's so convenient. Obtaining the finance to buy is relatively easy. Although there may be cheaper ways to borrow capital, hire purchase remains the firm favourite (see page 11). You, the dealer will not only get paid in full by the agreement company for the car, but should in addition receive a commission for selling the finance package.

A Consumer Credit Licence is required, enquire in the first instance to your local Trading Standards Office (number in the 'phone book). Names and addresses of H.P. Companies are listed at the end of the book [13].

DEALER AND TRADE INFORMATION

TRADE ASSOCIATIONS
For the sake of your reputation and protection, you should consider membership of a recognised trade organisation. Contact the Scottish Motor Trade Association Ltd. [14], The Retail Motor Industry Federation [15], or The Society of Motor Manufacturers and Traders [16]. They have a code of practice, and in the event of unresolved disputes with customers, they will offer an arbitration service.

MOTOR TRADER'S INSURANCE
Specialist insurance is necessary for car dealers and most related industries such as mobile mechanics, valet services, etc. **Your private motor policy will not be valid.**

How much your premium will cost is largely dependant on whether you are trading 'with' or 'without' proper business premises - the latter being more expensive.

A basic policy should cover you for public liability, including claims incurred by defective workmanship, or faulty goods, ie. spare parts, vehicles, etc. Find motor insurers in the address section at the end of the book [17].

TRADE (LICENCE) PLATES
Trade plates are temporary red and white registration marks, readily affixed to any vehicle. They enable a motor trader (or similar) to legally drive on the road an untaxed vehicle which is temporarily in his possession, provided it is used in the course of his business. This trade licence may run for 6 or 12 months.

Application form VLT 301 is available from your local Vehicle Registration Office*. The plates are not granted easily or automatically; applicants have to meet certain criteria. They are carefully vetted for suitability, and must declare previous convictions related to vehicle licensing. Any falsehood when applying renders the applicant to a heavy fine or up to two years imprisonment.

* Note: with intended V.R.O. closures, check with Department of Transport (number in telephone book) for an update in your area .

VEHICLE MILEAGE CHECK LTD.

Any car dealer, especially in the retail sector, has to protect his reputation. Unfortunately, the motor trade as a whole is still tarnished by the unscrupulous dealers who 'clock', or turn back vehicle mileage readings. Unfortunately, even honest traders fall foul of the law if they sell a car in all honesty, as 'genuine mileage', but is later found to be not the case.

Vehicle Mileage Check Ltd. [18], was founded by an ex Trading Standards Officer, and is the longest established mileage check company in Britain. Subscribers pay an initial enrolment and setting up fee, followed by an annual administration charge. In addition, enquiries are bought in advance, ie. blocks of 20, 30, 40, 50, or more.

When an enquiry is initiated by a subscriber, V.M.C. will undertake to establish if a recorded mileage is genuine. Even if their search proves inconclusive, the fact that you have made a serious attempt to establish the true mileage is recorded by V.M.C., and could prove beneficial to your defence in any subsequent court case related to mileage fraud.

Customer confidence and dealer image is boosted when the V.M.C. poster and certificate is publicly displayed - you will be perceived as being one of the more scrupulous dealers.

H P INFORMATION PLC

Since 1938, H.P.I. has been operating with the aim of combating fraud within the motor industry, by supplying relevant data to their members.

Upon receipt of an annual subscription (and a further nominal charge for each individual enquiry), HP Information [19] will if possible, divulge whether the car is *(a) the subject of a finance agreement, (b) reported as stolen, or (c) on the Vehicle Condition Alert Register,* ie. it has been an insurance write off. See page 71 for more information about V.C.A.R.

All enquiries are backed up with written confirmation.

For certain categories of subscriber, H.P.I. will divulge if a car has had a cherished registration mark transfer (applies from 1990 on).

The services listed above are only some that are available from HP Information, their databanks cover commercial vehicles, motorcycles - even aircraft!

ADVERTISING DISPLAY
Vital to any car retailer is professional presentation. Without advertising signs, headboards, stickers and even the dreaded bunting, life just wouldn't be the same! Contact Portfolio Marketing [20] for more details.

TOWING FRAMES & DOLLIES
Suitable for transporting cars to and from auction, these devices are ideal for one man operation - gone are the days of the cumbersome Land Rover and car transporter combination. The only modification necessary for your car being the simple addition of a towbar.

The *towing frame* is a 'V' or 'A ' shaped steel device which is attached to the suspension / steering gear of the towed vehicle. The latter is unsuspended, and because it rides on its own road wheels, is unbraked.

A *tow dolly* is a two wheeled trailer which cradles the (suspended) front wheels of the towed vehicle. Basic dollies may be unbraked, while the more expensive versions will be braked.

In the case of *unbraked* frames or dollies, there may be legal complications. It is possible that they may be used legally for recovery, ie. when the towed car has broken down. Apparently the law is unclear on this, so it may be a wise precaution to temporarily immobilise the vehicle as an extra safeguard.

Before employing these towing devices, check your insurance policy to ensure you have adequate cover.

Find stockists of these devices in the address section at the end of the book [21].

SPECIALISED CARS

COLLECTOR'S CARS
Until you have gained the necessary experience and expertise, it would be wise to keep clear of classic or exotic car dealing. Without this (and a high bank balance), you run the risk of (a) having capital tied up and holding stock that is difficult to turn over, (b) being 'conned' into buying cars in need of expensive repairs, and (c) finding it very hard to establish a car's true value.

For those who know what they are doing, this market can be very lucrative. Because dealing in exotics involves higher risk, a suitable entry level could be gained via 'classic' cars. These offer potentially high profit margins combined with less need for high business turnover. If anything, stock will appreciate as opposed to depreciate whilst in your care, and there is the added bonus of knowing that because your stock is unique, the customer has little chance of finding a similar vehicle elsewhere!

PRICING
We have no way of knowing what to-day's cars will be worth tomorrow. Classic car prices have tumbled after the over inflated prices of the 1980s. Fashion as well as time can create the 'cult' classic car - notable examples being the Morris Minor and the V.W. Beetle. Guides on pricing may be found in the specialist magazines, and in the comprehensive **Miller's Collector's Car Price Guide**, a professional handbook, available through bookshops. Details are also given for motorcycles and automobilia, and the guide touches on diverse items such as fire engines and children's pedal cars.

No price guide can be accurate in this market, as values depend on rarity, condition, and desirability.

Newer, so called classic cars bought from private individuals fluctuate wildly in price for various reasons. Supply and demand is one factor, the seller's perception of his car being another; one owner may see his car for what it really is - a rusty old banger. Another owner sees his (identical) vehicle as 'an appreciating classic', pricing it as such in his advert. Both cars sell, with the second car possibly achieving twice the price of the first.

Buying privately is probably your best option. You will have a good chance to inspect the vehicle (bearing in mind the high cost of potential repairs, and poor spare parts availability), and you may be able to drive a hard bargain when haggling.

You will have to be willing to travel for your cars. A good source for stock is through the 'Collectors and Classic Car' sections of the regional motor mart type magazines. Some of these will accept subscriptions, contact some based outwith your area [22], for maximum benefit.

Make the most from your stock - you could use your vehicles for film or advertising work, by registering with a specialist agency - see the *'Promotion and Agency Work'* section later in the book.

High demand at classic car auctions creates high bidding, and you may wish to sell stock by this method, having bought from private sources - a reversal of the usual way dealers operate! Time will have to be spent 'on site' to research prices. If buying from auction, check if you are subject to a buyer's premium.

Classics are sometimes imported from sunny climates, popular places being California and South Africa. These cars can be almost rust free. Should you be offered an example which has not yet been registered in the U.K., ensure it is supplied with Customs and Excise form 386, which is confirmation that import duty has been paid.

Malta is rife with British classics - all right hand drive, not much rust, but plenty of bashes!

NON RUNNERS, STOLEN / RECOVERED,
DAMAGED REPAIRABLES, AND THE V.C.A.R.

At the bottom end of the market, is the time honoured method of trading, namely buying cheap non runners, spending time and money on them to produce car which can be sold at a profit. Becoming more popular is the trend of fixing up accident damaged or stolen / recovered vehicles. This area is bordering on the shady area of the motor trade, so appropriate stock should be selected very carefully, and any work done must be carried out to a high standard.

Some vehicles may look as if they are only damaged slightly, but a twisted body may be hidden which could prove impossible to rectify.

Sources for these vehicles are; general auction sales, specialised salvage auctions and dealers, or direct from the insurance company with which the total loss was claimed.

Note: NEVER sell a vehicle that is anything less than totally safe.

If a vehicle was the subject of a substantial insurance claim, details are likely to have been lodged with HP Information's *Vehicle Condition Alert Register* (V.C.A.R.) - see page 68. H.P.I. are referred to by the trade to establish if a potential purchase has any undesirable history. The fact that a car is listed on V.C.A.R. may even be noted on its V5 registration document. Being on this trader's 'black list' will obviously reduce its value substantially. If the car is sold by auction, the auctioneer has a moral obligation to announce its inclusion on the V.C.A.R. register.

A vehicle when removed from V.C.A.R. will in theory, regain its market value.

Once properly repaired, it should be taken to an independent centre authorized by H.P.I., where stringent testing will be carried out. If the car reaches the required standard, the centre will notify H.P.I. who undertake to remove it from the register within three working days. These approved centres may also carry out the necessary repairs, and will deal with enquiries from both the trade and general public.

Two companies offering the above service are Popplewells of Essex [23], and Autolign Ltd. [24], who have depots at Glasgow, Belfast, Leeds, Gateshead, Manchester, Halesowen, Bristol, Redhill, Walthamstow, and Northampton.

CAR AUCTIONS

GENERAL INFORMATION

Buying and selling at auctions can be very profitable, but is not a pastime suited to the faint hearted, or those unwilling to take a risk. Self discipline and strong willpower are the order of the day!

Buying from auction offers wide choice coupled with genuine 'trade' prices, but purchasers' rights are limited. Aside from statutory entitlements, those that do exist will be laid out in the 'Conditions of Sale', which will be shown to you on request.

The buyer is subject to intense pressure; the car in the sale ring will be sold within a couple of minutes. There is the risk of buying a rogue car (with little or no warranty), and there is no possibility of a test drive. The description of the car as given by the auctioneer should be listened to intently. It is very easy, when caught up in the excitement, to pay more than was intended for a well presented car that has hidden faults. When buying any vehicle from an auction, you should expect and be prepared to spend some money on repairs to bring the car up to 'spec'.

Most auction houses charge a compulsory *indemnity fee* to the buyer. This charge is over and above your bid price; it will cover you in the event of the car being found out to be stolen, an insurance write off, the subject of a hire purchase agreement, or the mileage being misrepresented.

The seller at auction is almost guaranteed a quick sale with little chance of any 'comeback'. For a private seller this is good news - rather than trading in his car to a dealer, entering his car to auction gives him ready cash, which he can use for gaining a better discount. For the business seller, the sale releases instant cash to relieve his cash flow situation.

On the negative side, the seller will get a low market price for his car. Deducted from this will be a percentage commission plus an entry or administration fee. He, or his agent will have to take time out to deliver the car (although some auctions offer a pick up service), and possibly spend more time to be present for the duration of the sale.

See page 69 regarding tow frames and dollies if you intend to deal at auctions without the assistance of another person.

Yet again, the theme that runs throughout this book - do your homework! Attend as many sales as you can, whether buying or selling, before participating.

In the event of a dispute arising between customer and auction, a resolution may be found by contacting the Society of Motor Auctions [25], assuming the auction holds membership.

A list of auctions can be found in the address section [26]. Check your copy of Yellow Pages for addresses of sales locally.

AUCTIONS - BEST BUYS
The auction is a gamble. Everyone who sells a car has a reason for doing so. You will have to minimise the risks, and the following categories of car should prove to be safer bets:-

Ex - Mobility
These cars are leased or lent by the appropriate social authorities to infirm or physically disabled drivers over a set number of years. When this period expires, the vehicles are returned, then put to the trade.

The auctioneer will state that the car is ex-mobility. As it is designated as a different taxation class on the V5, the new owner will have to apply for a new tax disc, when the car will be changed to the usual PLG category.

Ex-mobility cars are as a rule good buys; they tend to be low mileage, driven by mature owners, and well maintained.

Ex - Government (Ministry of Defence / Public utility services)
Don't let the high mileages put you off. These cars or vans may have been driven hard by many different drivers, but should be well serviced, and had new parts fitted without any thought of cost! Public utility vans or cars may be fair, but ex police cars are usually the best buys in this category.

Ex - Rental
These cars may have had some abuse in their short lives, but will have been well maintained. Most rental companies renew their models after a set period (usually 6 or 12 months), as a matter of course. Look for batches bearing similar registration numbers (see below). Many originate from the Channel Islands, benefiting from a moderate climate and a geographically restricted area - no 300 mile thrashes down the motorway!

Entries bearing consecutive registration marks
Ex-fleet, similar to those listed above, or ex-company cars are likely to be reasonably safe buys, especially if they bear (reasonably) consecutive registration numbers, indicating that they have been entered *en masse* due to company policy, ie. age. It is unlikely that they will have been 'tarted up' by someone keen to make a quick profit.

Be more wary of the solitary vehicle entry, even from a business source - it may be presented for sale due to more serious reasons.

Ex - Manufacturer
Manufacturers sometimes register cars in their own name to boost their sales ratings. These are virtually new vehicles, despite having a previous keeper listed on the V5. They should be excellent value, coming complete with a full factory warranty.

Cars in need of aesthetic attention
Many traders look for sound cars with poor paintwork or appearance. After a quick respray, the car is re-entered to reap the benefits of their effort. Although this method of dealing is profitable for some, it doesn't do a lot for the well being of mankind. See the later section, *Trade Dealing Between Auctions*, on page 82.

BUYING CARS AT AUCTION - CHECKLIST
The following items should be helpful to you:-

    a) *Toolkit*
    b) *Towrope*
    c) *Petrol can (full)*
    d) *Set of jump leads*
    e) *Loose change or 'phonecard*
    f) *Insurance broker's telephone number*
    g) *Method of payment*
    h) *Car price guide*
    i) *Pen and paper*
    j) *Torch and magnet*

BEFORE THE SALE
With the above items, arrive one or two hours before the auction starts. You will then be able to take your time, inspecting cars of interest and taking relevant notes. You will have done the homework, and set your budget level.

If there happens to be an abundance of one particular model, theoretically, the last few sold should attract lower bids as the dealers will have purchased their quota for the day, and most private buyers will have bought their single purchase. Occasionally this happens, but in practice, dealers have unlimited funds when bidding for that 'steal', and if a dealer is in attendance, bang goes the theory! Don't expect to buy a car for nothing.

You will need luck to buy rarer cars at a good price - a private buyer will let his heart rule his head, while the dealer who specialises in that type or model will be willing to pay more than the opposition.

Assess the car before it enters the ring. Checking out a car has been described earlier in some detail (see pages 23 - 35), but when buying at auctions your inspection will be more restricted. Look for the signs associated with previous accident damage, or a recent respray, ie. missing badges and trim which haven't been replaced.

Uneven tyre wear should be checked (page 28), as should the signs of high mileage (page 21).

Look at the deposits inside the exhaust tailpipe; grey means (hopefully) good news. Black - a rich petrol mixture or engine wear, and white - either a lean mixture, or a long fast drive to the auction site! Watch out for excess water or oil deposits, this could mean future expense for the new owner.

Cars to avoid are those entered by small time traders. Clues are; the recent respray complete with coachlines, new accessories, such as wheel trims etc. The car will have been immaculately valeted, and the tyres probably treated with a rubberised dressing.

Look for clues that the car was used as a taxi or driving school car. Tell tale signs might be; the roof is scratched, holed, or there is differing shades of paint where a sign has been; traces of signwriting or holes on the rear bumper; marks inside the windscreen made by extra mirrors. If you can access the interior, look for evidence of taximeters or broadcasting equipment. Lift the carpet in the passenger's footwell near the bulkhead, are there holes or cuts in the carpet, where dual controls were fitted?

Certain auctions display an engineer's report on the windscreen of selected vehicles. Whilst this may be encouraging, don't take the findings as 'gospel'. Time and economic factors won't permit a thorough inspection of the sale car, so findings may be vague or not entirely accurate.

When the car you are interested in is started up to be driven into the sale ring, make sure you stand alongside the car, just behind the driver, so you can see the instrument panel. As the engine is turned on, the ignition and oil warning lights should immediately extinguish. Listen for any rattles or knocks which may disappear after the engine has run for a few seconds. As soon as you have established the above, turn your attention to the exhaust emission - look for any blue smoke or water vapour.

Doors are kept locked before the sale, but when the car is being driven up to the rostrum, the driver may pause, or even leave it unattended for a couple of minutes. Look inside the glove compartment for a service history book - worth its weight in gold if it has been dealer stamped and up to date. A real asset if you are buying with a view to resale.

Ask the driver to pull the bonnet release, now you have a chance to inspect the engine - ask the driver how the clutch feels, or if gear selection was easy. He may volunteer other useful information. Keep your wits about you, some auctions let the *owner* drive his own car through the ring!

AUCTIONEER'S 'SPEAK'
To the initiated, the auctioneer almost gives a running commentary on how the sale is progressing! Listen out for the following phrases (or similar):-

"There's a long way to go..." means the vehicle is attracting bids at well below the reserve price.

"I'm just wasting my time..." still not near the reserve price, he is threatening to withdraw the car from the sale.

"Take it away, driver..." the car is withdrawn.

"It's against the owners..." the offers are nearing the reserve.

"I'll try it for you...", "I'll give them a 'phone..." the final bid is near, but below the reserve price. The auctioneers will contact the vendor, and try to persuade him to accept the offer.

"I'm puting the car on sale...It's here to be sold...It's on the market...All your own bids now..." bids are above the reserve, and the final offer will be successful.

"For the last time, are you all finished...?" the vehicle is about to be sold on the drop of the hammer, or alternatively withdrawn from the sale.

BIDDING & BUYING

Traders can and do bid for their own cars until reaching the reserve price, and beyond, leaving the excited private buyer bidding for a car that is overpriced (see page 81).

The auctioneer will open the sale by inviting offers. He will imply what the opening bid figure should be. This figure will be pitched at the retail price, or just above. He will progressively reduce the suggested figure until he receives the first bid. Leave your own offer until as late as possible - try not to be the first bidder. Some less reputable auctioneers may create imaginary bids to boost the price.

When making your initial bid, establish eye contact with the auctioneer, and give a definite nod or wave. You can rest assured that after establishing contact, the auctioneer will seek you out for further offers, so a nod when bidding should then be sufficient. This way, you will remain anonymous to the seller of the car who could recognise you as a new face, and try to 'bid up' the price.

Keep calm, stick to the budget, and don't be pushed into making an impulse bid. Should you feel this kind of pressure, walk away, there will be many more opportunities later.

If successful, your documents will be supplied when you pick up the keys. Get hold of the V5, and as previously mentioned, contact the previous owner / keeper (note that he may not necessarily be the same person that entered the car to the sale), to establish relevant details and the genuine mileage. Should you find the mileage to be false, the description of the car inaccurate, or you find major faults within the allotted period of auction warranty, you should return the car to the sale, and take up your case with the auctioneers. If they accept your complaint as being valid, you will be offered the car at a suitably revised price, or have your money refunded in full. Should the car be found to have been stolen or subject to hire purchase, you will be covered by the auction's indemnity scheme.

Check that you are fully insured before driving away from the sale.

When the vehicle that you were interested in remains unsold, you might be able to negotiate with the seller, via the auctioneers, after the sale. One advantage of doing this is that you could ask for a test drive before making an offer!

PAYING

Unless your face is known to the auctioneer, you will have to pay a cash deposit on your purchase immediately on the fall of the hammer. This will be a reasonably substantial amount, as specified in the 'Conditions of Sale'. The balance will have to be paid in full before you take possession of the car keys. Obviously, the acceptable method of paying will be cash (be careful when carrying large sums of money), also banker's draft or auction account card, if applicable. Credit cards are not usually taken as payment, but business cheques will be accepted by prior arrangement with the auctioneer. Personal cheques may be accepted by arrangement, but you might not be allowed to remove the car until the cheque has received bank clearance, which could be a few days later.

DON'T GET CONNED - AUCTIONS

Readers are advised to refer to the previous article on page 36.

Be wary of any description label being affixed immediately prior to the car entering the sale ring; there may be something about the vehicle (high mileage, no M.O.T., etc.) that the seller would rather not advertise.

Listen to the auctioneer's description of the car very carefully, all may not be as appears - a 5 speed gearknob may have been fitted to a 4 speed car; 'Ghia', 'L', and 'injection' (or similar) badges may have been attached to more basic models.

Supposedly 'broken' bonnet release catches might be suspicious - maybe a close inspection of the engine compartment is not welcome!

Take note if the vehicle is being driven into the sale ring by auction staff. *The seller may prefer to drive the car in himself because he knows of an existing fault,* eg. there is no first gear, and he knows how to drive the car in a manner that would disguise the problem.

Unscrupulous buyers may try to purchase the car cheaper by putting the other bidders off. They may try swapping the ignition leads around to make the car engine run erratically - easily fixed once they've bought the car! When you are having a close look at the car before the sale, other potential buyers may make disparaging remarks about its condition - don't take them at face value, these critics might also be interested in bidding for the car.

Bidding Up

Having put reserve prices on the vehicles that they are trying to sell, some dealers will bid for their own cars. With the knowledge that their cars will not sell for less, the only thing at stake is the entry fee (which in any case is probably reduced for these regular customers).

The dealer and an accomplice will be present for the sale. To make their act convincing, they will wait until the auctioneer has reduced his 'opening bid' suggestion (see page 79) to a plausible level. The dealer and his friend will start rapidly bidding against each other, until they reach a figure just below the reserve. Any other person interested is easily caught up in the excitement, he wants the car but hasn't been given a chance to either bid or think! When the two dealers stop bidding, the third party makes his first bid - at or above the reserve price, and bingo, the dealers have made their profit. Any further bids by other interested buyers can be an added bonus to the dealers.

Unofficial Sellers

The seller of the car you have unsuccessfully bid for may approach you, with a view to selling the car direct. He noted your previous interest, and is now trying to by-pass the auction process. Beware, it is possible the car is 'bad' ie., stolen, unroadworthy etc. The seller may never have intended selling through the auction in the first place - he could have entered the car under a false name, setting an unrealistically high reserve figure. When the car is not sold he offers you the car at a low figure (possibly even lower than your bid), telling you that he doesn't want the trouble or expense of re-entering the car. You pay cash, get the dud car, and have no contact address for any comeback.

Outside the auction premises may be found parked cars offered for sale, the seller sitting in the vehicle, or standing near, waiting for prospective buyers. Dealing with these people is asking for trouble. They are usually unlicensed 'cowboy' dealers who will tell you any story to obtain a sale; again, you may have no contact address as their name is unlikely to coincide with that appearing on the V5. Cash will be insisted upon as payment - the cars will probably be old 'bangers', definitely in the cash category. Be wary, you will have limited consumer rights.

TRADE DEALING BETWEEN AUCTIONS

Profit can be made purely by buying and selling between auctions. Expensive cars are best to deal with, as price fluctuations between different sales can run into hundreds of pounds per car. To be successful takes a great deal of knowledge and skill. See the *Bidding Up* section (above). Potential traders will have to be prepared to risk and lose money in the event of the system not working for them.

Any geographical price differences may not justify travelling and transport expenses. However, the stories you may have heard about the dealer who bought a scruffy car from a sale, gave it a thorough valeting, and entered it to a different auction, making a big profit *are true* - I've seen it being done. However, these chance windfalls are the exception, rather than the rule.

Traders dealing at the cheaper end of the market will work on lower profit margins - buying cars from sales, respraying, valeting, and re-entering them. To many, this is their main source of income, but the amount of work involved has to be justified. Entry fees and commission cut further into profits, although rates may be negotiated with the auctioneers, provided they could expect plenty of repeat business from you.

Potential traders will have to spend a lot of time researching prices at various auctions before experimenting with one or two transactions to 'test the water'. Bear in mind that a lot of time can be wasted hanging around an auction with the possibility of no sale at the end of the day.

REFUNDS OF VEHICLE EXCISE DUTY

Around 90% of vehicles passing through auctions are sold without a current tax disc. Either the disc has expired, or more likely the vendor has surrendered it to the authorities for a refund.

When auctioned, the car will receive no less a bid because it isn't taxed, but you as a dealer can profit from this little extra cash.

Eligibility for a refund applies if the licence was bought by you, or passed on to you with the vehicle. The refund will be based only on each full calendar month remaining. It is necessary to ensure the tax disc is received by the authorities *before the end of the month preceding the period you are claiming for.*When surrendering a six monthly disc, you will only receive the monthly rate applicable to a twelve month disc.

Income received from this source must be declared to the Inland Revenue.

Application form V14 is available from Post Offices or Vehicle Registration Offices*, and posted to D.V.L.C. [27], or handed in to your local V.R.O.

SELLING AT AUCTIONS

At auction, the bid price is almost dependant on the visual appearance of the car. A buyer can only judge it s condition by sight and sound, as there is no chance of a test drive. Preparation for selling a car was dealt with earlier, see pages 38 - 40.

Make a point of being present for the sale, with the vehicle immaculately presented. Hand in the V5, M.O.T certificate, and tax disc if appropriate to the entry office. It is possible to enter the car without documentation, but if this is the case, the vehicle *might* be sold as untaxed, or without M.O.T., and this will be reflected in the bidding. The lack of a V5 is not so important, as the buyer will be able to obtain a duplicate from D.V.L.C., in the event of you being unable to forward the original.

*See note on page 67.

An entry form will have to be completed. Be honest when filling this in - it will be to your advantage in the long run. Answer truthfully any questions asked by the clerk. If unsure about the genuine mileage, then don't guarantee it, even if the previous owner misled you - it is still your responsibility, and the buck stops with you! Ensure there is no hire purchase outstanding on the car.

Decide on the reserve price you wish to set, and the class of warranty you will give, if any (see below). Don't forget to mention any fitted extras, or any other attractive selling points.

An entry fee will be payable (usually a flat rate) in addition to a percentage commission which is calculated from the final bid price. These deductions will be made from the auctioneer's cheque, which you will receive a few days later.

In the event of the car remaining unsold, the entry fee will be payable when you uplift your keys, unless you intend leaving the car at the auction site until the next sale date.

RESERVE PRICES

Prior to the sale, let the auctioneer know what is your lowest acceptable bid, ie. the reserve price. The auctioneer will be willing to offer valuation advice if necessary. Be realistic when setting your figure, remember you are selling to the trade. A reserve price which is comparable to that of a private deal may result in no sale. Some auctioneers will decline setting a reserve on vehicles over a certain age.

Should the vehicle fall just short of its reserve, you may be called to the rostrum, "Is the owner in the audience?" If you are not in attendance, the auctioneers may try to contact you by telephone, and will usually apply some gentle persuasion on you to accept the offer.

Provided your reserve figure was sensible, it might pay you to re-enter the car in the next sale (the auctioneer should be willing to store the car until then). Alternatively, you could put the car to a different auction where the bidders won't recognise the car from 'last week's sale'. Remember that an entry fee is due each time the car is put to auction.

The entry fee may be reduced or waived altogether by special arrangement with the auction's better clients. This will depend on the particular auction's policy, and the amount of business the dealer turns over.

ENTRY CLASSES / WARRANTIES

All auctions have differing 'Conditions of Entry, Sale, and Purchase'. It must be emphasised that you carefully study these before deciding to use an auction. As a very rough guide, some typical examples of vehicle entry categories are listed here:-

Group 'A' - *"Full Warranty"*, *"No Major Defects"*, or *"Good Mechanically"*, all imply varying degrees of guarantee.

Group 'B' - *"Running Order Only"*, implies that the car is guaranteed to drive under its own power out of the saleroom, and no more than that.

Group 'C' - *"Company Car Direct"*, *"Main Agent Direct"*, *"Trade Vehicle"*, *"Trade In Vehicle"*, possibly no guarantee, but are probably honest entries. Some reputable companies have a policy of putting all ex-fleet cars to auction. Main prestigious dealerships may enter all stock that is unsuitable for retailing from their showroom.

Group 'D' - *"No Warranty"*, *"No Guarantee"*, *"Sold As Seen"*, no guarantee is given or implied for cars in these categories.

The above examples are very loose, and of course, the wording may vary between different auctions. Warranties may only cover limited mechanical components, ie. engine, gearbox, differential, etc. Ancillaries or 'minor' components, such as clutch, electrics etc., may not be included - even by the best warranties. Any guarantee given will expire shortly after after the end of the sale, possibly as short as one hour, maybe as long as one day!

Structural or chassis warranties are generally not given on cars over a specific age. Recorded mileage will only be guaranteed if described as genuine by the auction or auctioneer.

If any fault is mentioned at the time of the sale by the auctioneer, there will be no complaint entertained, so remember to listen carefully.

When following any complaints procedure, refer first to the auction's terms of business. In most instances, there will be a requirement to have the car returned to the sale at the time of complaint; a telephone call may not be acceptable.

Having received no satisfaction, and a deadlock situation with the auctioneer ensues, you may wish to take the matter further (see page 74). As a last resort, legal action could be taken, but this can be an expensive exercise, with no guarantee of success.

LIQUIDATION AND SPECIAL AUCTIONS
Most larger auction houses run special sales - sports, classic, low mileage, or batches of vehicles direct from the manufacturer. Sales are also commissioned by government departments and large companies, check the local and national press for details of forthcoming sales, especially from government surplus stocks, Ministry of Defence, local authorities, and health authorities.

Bargains can be found at company liquidation auctions, the receiver of the failed company having instructed a sale of company assets. Often these will include motor vehicles which might be bought for a reasonable price, due to the fact that dealers may not be in attendance.

TRADE DEALING

Trade dealing is possibly a safer way to get started in the used car business. The idea is to buy part exchanges from main dealers, selling them on to the trade, via auction. Complaints are almost non existent, provided you are honest with the auctioneers regarding mileage, mechanical condition, etc. Trade dealing offers a relatively low capital outlay with few bad debts, and low business overheads.

Most main dealerships operate a policy whereby the cars that they retail must be of a certain standard as regards age, mileage, etc. If any traded in vehicle falls short of this standard, it will have to be disposed of. Unless these cars are converted into cash, the dealer's cashflow may be stretched unnecessarily. The easy answer for them is to put these vehicles to auction, but doing so means transport costs, staff tied up, with no certainty of a successful sale. An unsold car may need to be left on site until the next sale, possibly incurring further valeting fees.

Certain main dealers prefer to dispose of these part exchanges to trade brokers. They may get slightly less than auction prices, but they know that the vehicles will be disposed of with relatively little fuss.

The biggest snag in starting up as a trade dealer is the fact that most of the business in your area may be sewn up already by someone else. Dealers will tend to do business with someone they know and can trust. A lot of main dealers have stopped using trade brokers in recent years, either through bad experiences, or because they want a bigger slice of the cake, trying to realise more for their cars by starting retail 'trade in corners', etc.

Persistence, courtesy and patience may pay dividends, could strike lucky, and break in to an untapped market in your area.

Vitally important is your knowledge of current prices. Armed with your business card and price guide, make your offers below that which they should expect from auction - don't go above, otherwise you will defeat the object of the exercise! Experiment first, by buying only one or two vehicles. The cars that you are aiming for should be worth around £100 - £600 (maximum), minimising the chances of financial disaster.

After the introductions, you will be directed to the garage's back yard, where the humble trade ins are kept! You will want to make a thorough, but brief inspection of their condition. Hopefully, the dealer will volunteer information about known faults, and the price will be adjusted accordingly.

Offer cash - it's more likely to clinch the deal, and stick firmly to your offer.

As you will be selling the car through auction, minor mechanical faults should be of little concern to you. Should you be sold a rogue car, avoid complaining if possible. The car would have been sold to you in good faith, you may cut off a lucrative source for new stock, and in any case, the car was sold to you by trade, not retail - which limits your rights as a consumer.

Once the purchase is made, you will have to organise transport. The garage is unlikely to deliver, so the assistance of a friend will be desirable. Alternatively, you may wish to employ a one man operated towing frame or dolly, as referred to on page 69.

Your stock has to be presented well, so expert cosmetic treatment is called for. Cars are sold successfully at auctions by their appearance. Although you might have extra protection against comebacks, **you can still get into severe trouble for selling an unroadworthy vehicle.**

Refer to the previous section regarding auctions.

MOTOR TRADE ABBREVIATIONS

ABS	Anti lock braking system
AUTO	Automatic transmission
BHP	Brake horsepower
BRG	British racing green
CAT	Catalytic converter
CC	Cubic centimetres
C/L	Central locking system
CV	Constant velocity (joint)
DHC	Drophead coupe
DOHC	Double overhead cam
DVLA	Driver & Vehicle Licensing Agency
DVLC	Driver & Vehicle Licensing Centre
EFI	Electronic fuel injection
E/SR	Electric sunroof
E/W	Electric windows
FFSR	Factory fitted sunroof
FHC	Fixed head coupe
FSH	Full service history
FWD	Front wheel drive
GC	Good condition
GRO	Good running order
HP	Hire purchase
HPI	Hire Purchase Information
HRW	Heated rear window
K	Thousand
KM/H	Kilometres per hour
LHD	Left hand drive
LPG	Liquid petroleum gas
LSD	Limited slip differential
LWB	Long wheelbase
MOT	Ministry of Transport (certificate)
MPG	Miles per gallon
MPH	Miles per hour
NUCA	Nationwide Used Car Arbitration
O/D	Overdrive
OHC	Overhead cam
OHV	Overhead valve

ONO	Or nearest offer
O/O	Offers over
OTR	On the road (price)
OVNO	Or very near offer
PAS	Power assisted steering
PDI	Pre delivery inspection
PDM	Passenger door mirror
P/EX	Part exchange
PLG	Private Light Goods
P/PLATE	Private plate
P/X	Part exchange
RHD	Right hand drive
R/M	Recorded mileage
RWW	Rear wash wipe
S/R	Sunroof
S/S	Stainless steel
SWB	Short wheelbase
TD	Turbo diesel
TSO	Trading Standards Office(r)
T&T	Taxed and tested
UJ	Universal joint
V5	Vehicle registration document
VCAR	Vehicle Condition Alert Register
VGC	Very good condition
VIN	Vehicle Identification Number
VRM	Vehicle Registration Mark
VRO	Vehicle Registration Office
WHY?	What have you?

VEHICLE REGISTRATION MARKS - YEAR LETTERS

Suffixes

A 1963   B 1964   C 1965   D 1966

E ....   January to July, 1967

F ....   August 1967, to July, 1968

G ....   August 1968, to July, 1969

H ....   August 1969, to July, 1970

J ....   August 1970, to July, 1971

K ....   August 1971, to July, 1972

L ....   August 1972, to July, 1973

M ....   August 1973, to July, 1974

N ....   August 1974, to July, 1975

P ....   August 1975, to July, 1976

R ....   August 1976, to July, 1977

S ....   August 1977, to July, 1978

T ....   August 1978, to July, 1979

V ....   August 1979, to July, 1980

W ....   August 1980, to July, 1981

X ....   August 1981, to July, 1982

Y ....   August 1982, to July, 1983

VEHICLE REGISTRATION MARKS - YEAR LETTERS

Prefixes

A .... *August 1983, to July, 1984*

B .... *August 1984, to July, 1985*

C .... *August 1985, to July, 1986*

D .... *August 1986, to July, 1987*

E .... *August 1987, to July, 1988*

F .... *August 1988, to July, 1989*

G .... *August 1989, to July, 1990*

H .... *August 1990, to July, 1991*

J .... *August 1991, to July, 1992*

K .... *August 1992, to July, 1993*

L .... *August 1993, to July, 1994*

M .... *August 1994, to July, 1995*

N .... *August 1995, to July, 1996*

P .... *August 1996, to July, 1997*

R .... *August 1997, to July, 1998*

WORKING WITH CARS

ALL MOTORING SERVICES

FRANCHISES

The ideas promoted in this book may be attempted single handed, or as an individual working under a franchise scheme.

A *franchisee* will be self employed, but pays a parent company (the *franchiser*) for the benefits associated with that larger company, and may use the franchiser's name. Although technically a business in its own right, the franchise will be backed up by increased purchasing power, publicity, sales promotion, and service backup.

Franchise fees will vary between different companies, but may consist of, (a) an initial set up fee, (b) a regular subscription, and / or (c) a percentage of the takings.

When considering the franchise option, satisfy yourself of the franchiser's credibility. If the company happens to be a member of the British Franchise Association, this will be to your benefit. The B.F.A. operate a strict code of ethics, which participating member companies are obliged to observe.

A franchise information pack, an annual directory, and a franchise manual are published by the British Franchise Association [28], contact them for further details.

Note: *None of the companies mentioned within these pages are necessarily recommended by the publisher.*

When considering a franchise plan, take appropriate legal advice before making any commitment. Your solicitor could retain any monies until all transactions are completed to your full satisfaction.

DRIVING TUITION

Although you might not make a fortune as an *Approved Driving Instructor* (A.D.I.), the job does have its compensations. You can tailor the hours to suit yourself, working either full or part time. You will be your own boss, and have a unique sense of job satisfaction when you see a delighted pupil who has just passed the driving test. Conversely, you will probably work unsocial hours, including weekends, and suffer the frustration of difficult pupils. Nerves of steel are not required, but survival depends on skill, expert knowledge, and having a sympathetic disposition!

Strict requirements have to be met for entry and continued inclusion on the Register of Driving Instructors. Three rigorous exams have to be passed, and after qualification, the instructor is regularly assessed by the Department's examiners to establish 'continued fitness and ability to give instruction'.

Running a driving school can be cut throat, and many businesses fall by the wayside; price undercutting, running costs, and seasonal depressions all take their toll. However, compared with other ventures, capital outlay and overheads are relatively low, with few if any bad debt problems (lessons are paid for at the time, or in advance).

Contrary to popular opinion, an instructor won't 'hang on' to a pupil for as long as he can. Word soon gets around that he gives unnecessary lessons, and new business would soon dry up - an instructor's reputation is his best advertisement. The ideal instructor aims for a test pass as soon as is *safely* possible, using the qualities listed above; as well as showing patience, confidence, enthusiasm, and the ability to be a good communicator.

ELIGIBILITY
It is an offence to take money in return for driving tuition unless the instructor's name appears on the Register of Approved Driving Instructors, or he holds an official Trainee Licence.

To apply for inclusion in the register, you must be over 21 years of age, a full U.K. driving licence holder, having held the licence for at least 4 out of the preceding 6 years, with no periods of disqualification.

You should be a 'fit and proper person' - any convictions or motoring offences which are not spent under the Rehabilitation of Offenders Act, 1974 must be declared for consideration. It is possible for you to enter the Register if disabled, but you must be capable of using a manual gearbox during the relevant part of the qualifying examination.

THE QUALIFYING EXAMS
The exams fall into three parts, (1) the written test, (2) a test of driving ability, and (3) the test of instructional ability. The successful candidate will have to pay a new licence fee every four years, following expiry of the old one.

All three examinations have to be passed in order, and within two years of passing the first one; failure to do so means the candidate must start again, with the written test.

Only three attempts are allowed for tests two and three. If these are unsuccessful, the candidate will have to wait until the end of the two year period, before re-sitting all three exams.

You are unlikely to be successful in your attempts to pass exams two and three without first having some professional tuition. Try driving schools in your district (Yellow Pages) for recommendations of local instructor training establishments.

Be prepared to set aside at least 6 months to cover all exams; this is assuming that you manage to pass all three parts at the first attempt.

Part 1 - The Written Test
You will have ninety minutes to answer 100 multiple choice questions. The exam falls into four bands:-

a) Road procedure

b) Mechanical knowledge, car control, pedestrians, and traffic signs / signals

c) The driving test, law, and disabled drivers

d) Instructional techniques, and publications knowledge

A high pass mark is required at 85%. As well as achieving this overall figure, a score of 80% is necessary in each of the above categories.

The exam is carried out under classroom conditions at selected test centres throughout the country.

In my opinion, a training course is unnecessary to pass the written test, although a lot of commercial trainers would try to convince you otherwise. Around a quarter or more of the full training fee could be saved, provided the officially recommended reading is thoroughly studied and understood:-

The Highway Code (H.M.S.O. booklet)

DL25 Driving Test Report (form) *

Your Driving Test (H.M.S.O. booklet)

The Motor Vehicles (Driving Licences) Regs., 1987 *

The Motor Car - How It Works (a Ladybird book) *

The Driving Instructors Handbook, by J. Miller & M. Stacey (a Kogan Page book)

Items marked * are dealt with in the Driving Instructors Handbook.

You should also study part 1 (excluding section 4), and parts 2 and 4 of 'Instructional Techniques and Practice', by L. Walklin, published by Stanley Thornes Ltd.

Successful candidates will then be eligible to sit the second part of
the qualifying exam, the test of driving ability.

Part Two - The Test of Driving Ability & The Trainee Licence
The second part of the qualifying exam consists of an eyesight test; a
car number plate is to be read from a distance of 27.5 metres. Contact
lenses or spectacles may be worn, and only if the number is
successfully read will the driving test proceed.

The driving ability test is of an advanced standard; your chances of
success without having had professional tuition are low. You must
provide a suitable manual saloon or estate car.

The test lasts about one hour, and includes busy or fast moving
traffic, both inside and outside built up areas. There may be some
motorway driving involved. Brisk driving is required, keeping within
the speed limits. Manoeuvres similar to the 'L' test will be
scrutinised, and may include a reverse into a road on the right.

The result will be given at the end of the test, success entitles you
to sit part three, which is the test of instructional ability. At this
stage, candidates may apply for a *Trainee Licence*, which allows you to
accept payment for giving lessons. This certificate is valid for six
months, and is not renewable at the end of that period. The Trainee
Licence is optional, but strict rules apply with regard to its issue,
and as a working trainee, some supervision by another qualified A.D.I.
is mandatory.

Part Three - The Test of Instructional Ability
Probably the toughest of the exams, part three lasts around sixty
minutes (2 x half hour), with the Supervising Examiner (S.E.) acting
first as a learner in the early stages, and then as a pupil at test
standard. As before, you must provide a suitable car for the exam.

For each half of the test, the examiner will designate which subject
you as the instructor must teach, eg. crossroads, reverse to the left,
etc.

The exam is all role play, and it is quite unnerving to know that your 'pupil' is really assessing *you!* It is highly unlikely that you will pass this test without having been specifically trained beforehand.

The Supervising Examiner will be looking for:-

Correct and adequate instruction, given patiently, tactfully, and in a clear manner. Briefings given should not be too long, muddled, or repetitive.

Observation, analysis, and correction of the pupil's faults given without demoralising the pupil.

Instruction geared to the pupil's stage of learning or ability.

Upon passing this test, candidates are able to have their name entered on the Register, after payment of the necessary fee.

CHECK TESTS

After qualification, you will be required on occasion to take a test of 'continued ability and fitness to give instruction'. The S.E. will be present in your car, while you give a lesson to your pupil. Performance will be assessed, and after the lesson, he will offer constructive criticism and suggestions. The Supervising Examiner will then award you a grade between 1 - 6. This grading will also have a bearing on when your next check test will be. Low grades will have to be improved upon, and failure to do so within a specified time limit (or your refusal to take the test) will result in the removal of your name from the Register.

GENERAL

Further qualifications and diplomas are available, increasing job satisfaction, and enhancing your professional image.

Drumming up new business is often a struggle to the new instructor. Yellow Pages advertising is a must; you could offer a free lesson to any pupil who introduces new business to you. Also, you could have an arrangement with a local shop - organise free advertising by having the shopkeeper take bookings on your behalf, for which his commission could be the first lesson fee paid by the new pupil.

You may wish to concentrate on intensive, or residential courses. However, for a new driver, these generally produce less effective results than the same amount of tuition given over a longer period. There may also be a problem with your existing regular pupils as regards the mutual agreement of appointment times.

Alternatively you might consider franchising; in return for a regular fee, the company will supply you with a tuition vehicle, and pass on new customers, while you operate under the franchiser's name. Some companies will also train you, and might even organise a trainee licence.

Probably the biggest success in the franchised market is B.S.M. - the British School of Motoring [29], who can be contacted for more information.

Trade organisations, such as the Driving Instructor's Association [30], or the Motor Schools Association of Great Britain [31], exist for the benefit of instructors. Representations on behalf of the industry are made to government departments, and association members receive regular newsletters and updates.

R.C.M. Marketing [32] are suppliers of accessories, headboards, and teaching aids, etc., while He-Man equipment [33] will supply quality dual controls.

If you are interested in becoming an A.D.I., write or telephone the Driving Standards Agency [34], for their A.D.I. starter pack (available for a nominal fee). Relevant information and application forms are included in the pack.

CHERISHED CAR NUMBER DEALING

VEHICLE REGISTRATION MARKS

Generally, a Vehicle Registration Mark (V.R.M.) gives information of where and when a vehicle was first registered in this country.

From early this century until 1963, these marks consisted of letter(s) followed by number(s), then vice versa. These registration marks are now much sought after by dealers and collectors. Since 1963, year identification letters have been added to the (then) combination of six characters, so all present day marks consist of three letters, three numbers, and a year letter. For examples and identification tables, see pages 91 - 92.

These suffix and prefix derivatives are less sought after by dealers and collectors, unless they are of particular significance, or they make up a word, ie. COM 1C, MAG 1C, L1 LAC, etc.

The licensing authorities realised there was big money to be made from car numbers, and from 1st. August 1983, low (1-20) or interesting numbers were not issued as a matter of policy. These marks were retained with a view to possible sale at a later date by the Driver and Vehicle Licensing Agency (D.V.L.A.).

With the exception of marks sold by the D.V.L.A., the place of first registration is identified by the last two letters in the sequence, eg. ABC 123D - the letters BC indicate this (1966) car came originally from Leicester. BB 1234 was registered at Newcastle upon Tyne, pre 1963. In the case of even older cars, the only letter provides the identification eg., C 12 came from the West Riding of Yorkshire.

This system was operated by county and burgh authorities until 1974 with computerisation and the opening of the Driver and Vehicle Licensing Centre (D.V.L.C.) at Swansea. Some of these local registration letters were re-allocated to different areas via local Vehicle Registration Offices (V.R.O.s.). Thus two slightly differing identification tables would be needed - pre and post 1974, to accurately gauge an original registration authority. There have been many books written on this subject, most are now out of print, but a good source for these is your local library.

HOW TO TRANSFER A REGISTRATION MARK

Obtain a form V317 from your local Vehicle Registration Office. To get their address see leaflet V100, available from Post Offices, or look in the telephone directory under Department of Transport.

Vehicles involved in the transfer must be of a class which is subject to M.O.T. testing (this excludes vehicles such as milk floats, agricultural machines etc.).

It is possible to transfer numbers from motorcycles and mopeds to other vehicles (including motorcycles and mopeds), but not *from* other vehicle categories *to* motorcycles and mopeds.

It is not possible to transfer a number bearing a late year letter to an older vehicle, thus making it appear newer.

Both vehicles don't necessarily have to be registered on the V5 with the same keeper's name.

If a vehicle bearing a cherished number has been stolen, it may still be possible to salvage the registration mark, provided the car has been recorded with D.V.L.C. for over one year as such.

Hand in the completed application to the V.R.O. If you are lucky, the transfer will be completed while you wait, but in most cases the process should not take more than a few days.

You should enclose:-

> *The two V5 documents*
> *Two M.O.T. certificates*
> *Two current tax discs*
> *The transfer fee*

If the *donor* vehicle's tax has expired, this will be acceptable provided the expiry date is not more than six months prior to the application - tax discs should not be sent with postal applications.

Officials may wish to inspect one or both vehicles to establish their authenticity, by checking V.I.N. or engine numbers.

If there is no recipient vehicle available, it may be possible to retain the number for a period of up to 12 months (extendable). Apply on form V778/1, available from your V.R.O. In all cases, upon successful transfer, the newly affixed plates must have their characters properly displayed and spaced as regulations dictate.

THE DEALERS

Most car number dealers operate as agencies. Their advertisement lists appear in the Sunday newspapers, *Exchange & Mart*, and motor market type magazines. Using form V317, they will organise the transfer by telephone and post, without buyer or seller having any communication with each other, all transactions being via the agent.

Dealers may offer to search for a particular set of initials for a client. They will place a 'wanted' advertisement in a local newspaper serving the district of the original issuing authority (see page 101).

Sometimes the agent will actually own the vehicle that bears the number he is selling. He may have bought the vehicle cheaply by placing an advert in a car or motorcycle magazine. A scrap motorbike, or even a bare frame may be bought and rebuilt relatively cheaply, provided the number justifies the effort - the machine can then be sold after the transfer. If the agent is V.A.T. registered, he has to charge tax on the sale of the number.

All V5 s should be checked before going ahead with any purchase; the registration mark may be designated 'NON TRANSFERABLE'.

REGISTRATION MARK VALUES

Number plate values depend on availability, demand, and desirability. Rarc, namc, novelty, and very old numbers are at the top of the list of sought after plates. Another rule of thumb (generally) being the shorter the V.R.M., the more it is worth.

Homework has to be done, it is a relatively simple task to compile your own price guide. Reference books on the subject will help establish the existence of particular numbers, and their date of issue. Scour press advertisements for similar combinations on offer by other dealers, and study prices.

It is interesting to see the same registration mark being sold by different agencies - the seller using the services of more than one dealer. You will notice a wide variation in asking prices.

When setting up as an agent, try to keep profit margins low. There is a lot of competition out there; don't be shy in asking the seller what figure he has been offered by other dealers!

YEAR LETTER DISGUISES
Some owners prefer not to advertise the car's age with the registration year letter. These people are prepared to pay for an anonymous pre 1963 mark, or a suffix / prefix mark which is obviously a lot older than their car. Northern Ireland authorities still issue V.R.M.s that have no year prefix, but they are instantly recognisable by the letters 'I' or 'Z' being incorporated. These plates are legal in the U.K.

Dealers residing in Northern Ireland have a ready made export trade with mainland Britain - and an inexhaustible supply of dateless registration marks!

D.V.L.A. SALES
The Driver and Vehicle Licensing Agency offers direct to the public and dealers three different categories of sale:-

Custom Marks
Older prefix marks, ie. A, H, J, etc. Any combination of the three letters is possible, combined with numbers 1 - 20.

Upon payment (credit card or cheque), a certificate of entitlement is issued to the buyer, who must assign the new number to a vehicle within one year of purchase. Two pricing bands exist, dependant on the desirability of the mark. First come first served!

Select Registrations
These marks are allocated to new vehicles and can be reserved some weeks before their August 1st. issue.

Any available three letter combination is offered with numbers 1 - 20, also any number ending in '0' or '00'. Repeated digits, eg. 33 or 888 can be bought (but not 666). There are two pricing bands, dependant on the desirability of the mark. Again first come, first served!

The Classic Collection
Auctions are held of exclusive V.R.M.s (including pre 1963 marks) which have not been previously issued. Sales are held at venues throughout the country, and are advertised in the national press.

Telephone bids may be accepted, making personal attendance unnecessary.

Purchasers of any marks offered by D.V.L.A. are subject to an assignment (transfer) fee, in addition to the buying price.

For further information on sales, telephone the D.V.L.A. 'Hotline' [35], as given in the address section at the end of this book.

DEALER SERVICES
Many agencies charge unwitting clients for services which are offered by the Department of Transport cheaply, or even free of charge.

Some of these services are:-

Supply of D.V.L.A. Custom, Select, & Classic Marks
Some car owners still buy D.V.L.A. registration numbers through dealers, when they could buy direct - all it takes is a 'phone call to the Department's 'Hotline'.

Certain dealers buy up complete series of attractive combinations direct from D.V.L.A, thus creating a monopoly - forcing clients to pay a dealer's price, rather than that originally set by D.V.L.A.

Removal of 'Q' Prefix Marks
'Q' prefix marks have been allocated since August, 1983 to vehicles of uncertain age or origin; usually kit built, rebuilt, or certain imported models. These 'Q' letters are not age related, and were originally designated non transferable. 'Q' marks are not popular with car enthusiasts - especially classic car owners who would prefer a non suffix or prefix mark.

With a relaxation of regulations, these 'Q' prefixes may be removed if sufficient evidence of the vehicle's age can be provided. A replacement age related V.R.M. will be issued **free of charge.**

Re registering An Old Vehicle Under Its Original V.R.M.
If any elderly vehicle had not been issued with a computerised registration document (V5) by 1983, it was unlikely to be allowed to display its original registration number, and was possibly issued with an out of place 'A' suffix mark.

In 1990, the rules were relaxed, and subject to certain conditions, these vehicles were permitted to display their original V.R.M., on a non transferable basis.

Form V765 (available from V.R.O.s), details the documentary evidence required to prove the applicant's claim to the number, namely some or all of the following:-

> *The old 'log book' RF60/ VE60*

> *A recent photograph of the car*

> *Any other documentary evidence, eg. old newspaper cuttings, tax discs, M.O.T.s, etc.*

> *Completed form V55/5, or V5 if the car is currently registered under a different V.R.M.*

Your local V.R.O. will supply you with form V765/1, which is a list of officially recognised car / motorcycle clubs. The application should be initially sent to a club appropriate to your vehicle, for their authentication. You may be charged for this service.

Should you be successful, the club will return your documents, and you will then be eligible to register your vehicle at a V.R.O. If the car already has a different registration number, the documents will be sent to D.V.L.C., for their records to be amended.

Despite the fact that there is **no charge** made for re-registering a car by the D.V.L.A., some car number dealers will ask for a fee to carry out the same service.

Re - registering an old vehicle with an age related mark
Pre 1963 cars which were the subject of a cherished number transfer were often issued with an 'A' or 'B' registration suffix as a replacement. Recently, a **free service** has been introduced by D.V.L.A. whereby these vehicles may be allocated a previously unissued pre 1963 V.R.M., which would be in keeping with the car's age, since it bears no year letter. This number will be non transferable.

All that has to be done is to present your V.R.O. with the V5 document, M.O.T., and tax disc, for documentary amendments and the allocation of the new number.

Sections of the car number trade are taking advantage of this free service, and charging relatively high prices for doing so.

C.N.D.A.
The Cherished Numbers Dealers Association [36] is a member of the Retail Motor Industry Federation, and it represents reputable car number agencies, operating within their code of conduct. If you are buying privately, look for their logo in any advert. C.N.D.A. will provide advice and information on number buying, and offer an independent valuation service.

V.R.M.s FOR INVESTMENT
Interesting, or pre '63 marks can prove a good investment. Mine was originally bought for £500, and I was offered £1,500 for it six years later! If the original £500 had been deposited with a building society (receiving a generous 10% p.a. interest rate), the total after those six years would be £887 - a figure £613 less than the offer!

It is possible to insure a specific Vehicle Registration Mark with a specialist broker [37].

TAXI SERVICES

Potential licensees should note that rules and regulations governing all forms of taxi driving vary considerably throughout the country, depending on the policies of local administrations. The information that follows should be treated only as a rough guide.

All persons considering employment in this industry should be in possession of a clean driving licence, although it may be worthwhile checking with the licensing authority if you have any 'penalty points'. Declaration of any recent criminal convictions will have to be made, but all applications will be considered on their own merit.

HACKNEY CARRIAGES (TAXIS)
Hackney carriage driving involves hard work coupled with unsociable hours. There will be long periods of boredom. You should have the ability to deal with all kinds of people.

Cabs are allowed to solicit for business, and are hailed in the street or hired from the taxi rank. The total amount of working hackneys will be limited by the licensing authority. Customer charges are displayed by a meter fitted in the vehicle. Often there is a flat rate standing fee, followed by a mileage charge. There may be an additional charge if the taxi travels outwith its working area, ie. city boundary etc. The tariff is usually set by the local council.

Vehicles employed as hackneys may have to meet specifications laid down by the licensing authority. This is the reason that only the famous 'black cabs' (FX4 and Metrocabs) are used in certain areas. Random checks of the vehicles' condition may be carried out regularly by police, council officials, or one of the national motoring organisations.

A good income can be made - extras include tips - but insurance premiums are high. Buying a new cab is expensive; many licence holders prefer to work for an employer. Alternatively the cab can be leased or rented from the manufacturers via an agent. Contact your local Taxi Owners Association (see Yellow Pages) for further details.

Topographical Exam (The 'Knowledge')
To become a cab driver in most British cities, not only do you have to convince the licensing authorities of your good record and character, but fulfil various other requirements. You will probably have to meet set standards of age and medical fitness. Potential London cabbies may apply for a licence upon attaining the age of 20 years and 3 months, although the licence won't be issued to them until they reach 21.

You also may need to sit a special driving test, and pass the famous 'knowledge ' test. This topographical exam is hard work! It requires detailed knowledge of shortest routes between point 'A' to point 'B', street names, and location of public buildings, etc. Training can take up to two years, and many candidates drop out after just a few weeks. Mopeds are often used for route learning.

In provincial areas, this 'knowledge' may not be a mandatory requirement for the issue of a licence.

For licence application details in the Greater London Area, contact the Public Carriage Office [38]. Outside London try the National Federation of Taxi Cab Associates [39], or the licensing department of your local council.

PRIVATE HIRE
Minicabs, as they are sometimes known, are usually standard saloon cars, as opposed to the black cab. Licensed by the local council, they are subject to the 'for hire or reward' insurance category.

This type of taxi is pre booked, stopping for hires hailed in the street may be against regulations. The job rate may be a flat rate negotiated before the booking is made. The 'knowledge' may not be required, but vehicles will still be subject to regular or random checks by enforcement officers.

The licence may last for one year or longer, and the fee is payable to the local council.

A private hire firm may be called upon to deliver goods as well as passengers, ie. small parcels, oxygen cylinders to hospitals, etc., and may even win a contract to do so!

Some private hire companies employ their own drivers, whilst others simply act as booking agencies for the self employed, with a contract binding between them and the owner driver. Terms of contracts will obviously vary, so check the small print carefully before signing!

The company may charge the owner driver a flat weekly rate. All takings go to the driver, out of which he will have to pay his running expenses. As most minicabs are radio controlled, the company may charge rental on this equipment.

Applications for a private hire licence should be made to the relevant department of your local council.

CHAUFFEUR HIRE
Although an extension of private hire, in most cases the stringent regulations that apply to hackneys and even minicabs might not apply. However the cars must be covered by similar insurance which can be a major expense, although it is a good idea for newcomers to insure their vehicles on a short term basis, even for each separate job. This way the 'water can be tested' without having tied up much needed capital.

Rolls Royce, Bentley, and Jaguar / Daimler models of a certain vintage can be bought surprisingly cheaply - but must be well vetted before purchase, repairs can be cripplingly expensive. A private number plate not only adds a nice touch to these cars, but will also disguise their age.

Lots of market research is necessary before investing money in this kind of project; if there is a lot of competition in your area the idea may not be viable - resorting to price cutting can be very demoralising.

A high standard of (smooth) driving will be necessary, as well as a clean driving licence, and an even cleaner car! Although you should earn a higher rate per working hour than taxis, you will have to canvas for business, and the work that you do get will be irregular and spasmodic. Ensure jobs are paid for in advance.

Wedding hire is an obvious choice; approach established photographers for leads - they may be willing to display your advertising. Hotels may also be a good source for business - leave your colour brochures in the foyer. It is worth offering a commission for any bookings they take on your behalf.

Other services offered might range from executive businessmens' transport and simple airport transfers, to recreational work - country wide sightseeing tours and visits to places of interest.

Approach your local council to establish licence requirements for both you and your vehicle. If intending to operate from home, check first with the planning department.

COURIER SERVICES

Should you wish to start up your own courier service, you are well advised to get some 'hands on' experience by working first for a well established operation. You could be an employee, but more likely to use a vehicle of your own, on a self employed basis.

When choosing your vehicle, aim for economical models - diesel cars are ideal (motorcycles should have a *minimum* capacity of 125-150cc., unless being used purely for city centre work). Proper insurance cover will have to be arranged, and you will probably have to hire two way radio equipment or a pager from the courier company, for a nominal fee. You will be liable for your own expenses, ie. fuel, repairs etc.

Good local knowledge is necessary, with up to date atlases and street maps! You should be prepared to work any time of the day or night. The company also may require you to be on 'standby', which means in effect that after working a full day, you could be called upon at short notice to travel anywhere.

As with taxis, good money can be earned - when you get the work. Pocketing 50% of the charges is not unusual, in addition to any waiting time penalties. This penalty is levied on customers who haven't managed to get the package ready for dispatch upon arrival of the courier. Up to £1 for every five minutes that the driver has to wait may be charged.

Base yourself in the city if you intend working independently. A minimum of two people will be necessary. Publicity, advertising, getting your services known, and *remembered*, is of paramount importance. Call on as many offices and companies as possible, leaving your business card and introductory letter. Better still, desk top advertising gimmicks such as pens, ashtrays, and diaries will constantly remind potential customers of your services by keeping your name to the fore.

Refer to a copy of Yellow Pages for names and 'phone numbers of local courier services.

MOBILE VALETING SERVICE

Car owners appreciate clean vehicles but don't always like the work involved in keeping them that way! Subsequently, the mobile valeting market has grown in recent years. Nevertheless, people are only lazy to a certain point - if the job is too expensive they will do it themselves.

Hopefully, the biggest proportion of your turnover would be gained from repeat business. Private car owners could have the service carried out at their home or workplace at a specific time each week.

Commercial work is good business. Fleet cars and delivery vans need cleaning every week, eg. every Friday afternoon. Tuition cars at large driving schools may require a wash each day, or a regular visit to car auction premises on sale days might pay dividends.

The workplace is an excellent area for generating new business. Speak to the 'boss', ask permission to canvas the workforce during break or lunch hour. At least leave some advertising leaflets. Utilize other methods for publicity; 'fliers' tucked under windscreen wipers in car parks, your card displayed in local shop windows, and posters in shopping malls. Use local newspapers for advertising. When placing an advert in a paper which covers a large area, denote the district that you cover; this will save unnecessary enquiries.

Include a tariff on any advertising 'flier'. Offer discounts on regular business. Services will range from a simple car wash by hand, to a full valet.

Customers are always willing to pay a bit extra for the hand wash - machines tend to miss patches, and are rough on car paintwork. To maximise this service, and as an alternative to working mobile, you could rent disused premises, such as an old petrol station (even if only on a temporary basis to test the market). Choose a site in a busy area to attract passing trade.

Ensure your insurance covers you for accidental damage liability. Before any job, take written note of any body damage (use a line drawn diagram / plan of the car), and get the customer to agree and sign the sheet. This should help you in the event of any subsequent claim against you.

CHECKLIST
Obviously you will need the use of a medium / small van, or if you are not in a position to lay out capital, a car trailer. The following list may be of assistance:-

Sponge, chamois & bucket, cloths, scouring pads, lint free rags, toothbrush, hand brush & tray, small paintbrushes.

Pressure washer and tank, or 25 litre containers (for own water supply), garden hose and flexi brush with adapters for all tap fittings. Buckets, 12 volt vacuum cleaner, and 10",12", 13", 14", and 15" wooden, or plastic discs.

Silicone dashboard cleaner, upholstery, carpet, rubber / plastic trim, and glass cleaners, chrome and metal polish, detergent, wax wash additives, wax car polish, paint colour restorers and rubbing compound, rubberised tyre dressing, tar and fly remover, and possibly a range of touch in paints.

Overalls, rubber boots, plastic gloves, towels - spare clothes!

Mobile pressure washers are used with their own generators, and a water tank will need to be installed in the van, necessitating a considerable investment. Use Yellow Pages to find a local supplier. The wooden discs referred to above may be used to mask car wheels when applying tyre dressing.

'Autoglym' [40] are leading suppliers of car valeting consumables.

MOBILE VALET FRANCHISES
Autosheen [41], who operate the largest service in the U.K., offer a franchise to those who would like to start up their own business. With a full back up service, franchisees may work day and night, seven days a week. Although they may start out by cleaning cars themselves, franchisees are encouraged to progress by employing others, which leaves them free to manage their own business, and canvas for new trade.

WINDSCREEN CLEANING

An idea imported from the United States, screen cleaning has taken off
in this country, especially in London and other cities. The method
employed is dangerous, and budding entrepreneurs may find themselves in
breach of the law.

With client's permission, two people armed with a sponge, squeegee and
bucket attempt to complete the exercise in around fifty seconds, with
a car that has stopped at a red traffic light. The gratuity is usually
at the customer's discretion, but reasonable earnings can be made.

A safer, and more legal idea would be to offer a similar service at a
more appropriate site. Negotiate with the owner of a private petrol
station; he may welcome an unusual gimmick which keeps him one step
ahead of his competitors.

FAST FIT INSTALLATIONS

It is assumed that before attempting to put any of the following ideas into practice, you have some basic mechanical knowledge. You must be competent, and have the correct tools to suit any task. All jobs should be backed up with 'a money back guarantee', which instills customer confidence. Ensure you are covered with the appropriate *motor trade* insurance. Remember everyone makes mistakes, and you may be sued for something which is not necessarily your fault, or is beyond your control.

While each individual idea may not constitute a full time occupation, a combination of services should prove viable. If working on a part time basis, turnover will probably be low to begin with. As you will want to attract local trade, and newspaper advertising is relatively expensive, cheap adverts placed in shop windows might be effective. For any local tradesman, word of mouth and reputation are the best adverts.

Motor trade cash and carries will supply working stock, but potential patrons must prove that they are bona fide traders, by supplying the necessary references. A leading wholesale chain is Maccess [42], who have depots throughout the country.

SIMPLE SERVICING AND REPAIRS
Major vehicle repairs are best left to those suitably qualified and experienced. However there is always a gap in the market for routine maintenance, which doesn't require a high degree of expertise, provided the work is carried out properly, and in accordance with manufacturer's specifications. A 24 hour (overnight) service will be of interest to customers who must have use of their car every day.

A reasonably competent person should be capable of carrying out minor repair work. Vehicle workshop manuals (such as those produced by the Haynes Publishing Group) are available from retail outlets, and most contemporary models are covered.

Replacement part reference books and charts are available through wholesalers, motor factors, or direct from component manufacturers.

Some general data books available are:-

AUTODATA - *"Technical Data"* - Diagnostic testing, timing, routine servicing and mechanical checks are included for British, European, Japanese cars and light commercials.

HAYNES - *"Automotive Technical Data Book"* offers similar information, while a more compact book is published by *HAMLYN* - *"Car Service Data"*.

For repairs *QC PUBLICATIONS* [43] offers the trade their *"Parts Price Guide"*, which lists over 40,000 prices, and is updated quarterly. Also available is the *"Repair Times Guide"*, from which manufacturers' repair times can be estimated.

MINOR BODYWORK REPAIRS
Small localised bodywork repairs may be sometimes called for. Aerosol paints are available on the market, but results can often 'stick out like a sore thumb'. Holts 'Spraymatch' centres will supply a 300ml. spray can, mixed while you wait, matched to existing (even faded) paint. Contact the Bodycare helpline [44] for your nearest centre.

TOWBAR FITTING
Provided fitting instructions are followed properly, towbar fitting requires more time than skill. Being a lucrative sideline come spring and early summer, little capital outlay and low overheads make this service an attractive option.

Cheap towbars are always available, but a leading brand name such as Witter [45], means a superior product conforming to British Standards AU 113 and AU 114, and towbars are available for most vehicles.

SUNROOFS
These are relatively easy to install, and with the exception of an electric jigsaw, require little in the way of special tools. Stocks are available from trade cash and carrys. Be sure you are competent before hacking in to someones pride and joy!

I.C.E. AND ALARM INSTALLATIONS
Not to be attempted by the unskilled! Stick with one particular brand of product; familiarity means shorter fitting times.

Installing In Car Entertainment or alarm systems will involve some dismantling or trim stripping which can lead to complications - all cars are different.

Obtain stock from the cash and carry - you are able to make a profit on both product *and* service. Compile an illustrated product range portfolio, enabling the customer a choice of model before purchase.

EXHAUST FITTING (& Part Worn Tyres)
Many specialist exhaust fitting centres market their products as 'fitted free'. Try telephoning one of these retail outlets for the price of a full exhaust, then compare it with the 'take away' *trade* price asked by a good motor factor. Given that a lot of systems are fitted to the car by no more than three or four clamps, you will appreciate that high profits can be made by fitters.

Most 'complete' systems *exclude* the manifold, which customers may have to buy via the car main dealership. Difficulty is often experienced in the removal of old, rusted systems, so burning and cutting equipment will be necessary. Car ramps or stands are obviously needed.

A mobile service may be worth consideration, but unless there was an incentive such as incredibly low prices, the customer would probably find it an easier option to drive in to a fitting centre, without being bound by appointments.

Some years ago, there was a thriving market in 'part worn' tyres. They were imported from countries abroad where the minimum tyre tread depth was more than that of the U.K. With the raising of the minimum legal depth in this country from 1mm. to 1.6mm., these tyres were less attractive to the motoring public; sometimes the tyre would offer only about 0.4mm. of usable tread. With moves afoot to outlaw the sale (by dealers) of used tyres with a tread depth of less than 2mm., and the expense of fitting equipment, the business viability does not appear very encouraging!

WINDSCREEN REPAIRS (& Security Etching)
Coupled with ever increasing windscreen replacement prices, and stricter M.O.T. regulations regarding screens, the trend to repair rather than replace, is spawning a growing industry.

Thanks to modern technology, chipped or scratched windscreens may be repaired to a high standard. Kits are available which comprise the necessary resins, pressure pump, and ultra violet light gun. These kits are not suitable for repairing toughened glass windscreens.

The client may not need to pay for this service, as his insurance company will probably be billed directly - he might not even have an insurance excess to pay.

Franchise companies may offer training courses, and sales / technical back up. Refer to the address section at the end of this book [46]. Small 'do it yourself' kits are now available through retailers; it would therefore be wise to check out the potential market before making any business commitment.

To make a car less attractive to thieves, its registration number may be engraved on all the glass. Proprietary kits are available from car accessory shops and motor factors. Two methods of etching are used; either by applying a chemical over a stencil, which eats into the glass surface (no specialist tools necessary), or sandblasting, which is carried out with the help of a small compressor.

With the advent of D.I.Y. kits, the window engraving market has slumped, but although demand may be low, etching could be offered as a free incentive to customers considering any of the other services that you have to offer.

MOBILE CAR TUNING SERVICE

Because of the high capital outlay and technical nature of computerised car tuning, it is assumed that the sole trader who wishes to set up a mobile service would be making use of a franchise scheme.

Proper business premises will not be required, but a reliable van will be used to travel to jobs (invariably at the customers' home or place of work). The van may be leased or bought new - a finance package might be made available through the franchise company.

A second hand van may be acceptable, but will probably have to meet specific criteria set by the franchise company, ie. colour, type, make, age, etc. A diagnostic and tuning machine will also need to be purchased, probably by finance through the company, but most of the other equipment such as tools, initial stock, mobile 'phone, stationary, etc., should be supplied as part of the franchise package.

High earnings are possible, dependant on the effort and dedication of the franchisee. Previous related experience may not be necessary, but some mechanical knowledge would be advantageous. A residential training course may be offered, along with 'hands on' practice. Subjects covered will include routine servicing, engine tuning and diagnostics - possibly even car alarms, vehicle inspections, and caravan servicing! Franchisees should benefit from full technical back up and retraining, as well as regular information updates.

Some franchise companies will let you expand your business by taking on sub operators, or you could be allowed to sell out for a profit, provided a stipulated period has been worked successfully.

As start up costs are not particularly modest, a bank loan could be considered. The company may help your request by offering a business viability projection.

A one off payment is required for franchise name and rights. V.A.T. is payable, but may be reclaimed. In addition, a percentage of earnings will be paid to the franchisers. You will also need working capital for incidental expenses and replacing stock, etc.

Two market leaders in mobile tuning are *Hometune* (longest established, formed in 1968) [47], and *Computa Tune* (full British Franchise Association members) [48]. Contact them to see if any vacancies exist near you, or in other areas; this being the case, an interview may be arranged.

CAR SPARES

Millions of pounds are spent every year in the U.K. on new and used automotive spare parts. The capital required to set up a motor factoring business is beyond the average man in the street, without financial backing, however here are some ideas that could be tried on a part or full time basis.

In theory, an express service could be offered whereby specific orders could be uplifted from factors or dealers and delivered direct to the company or individual; the goods being charged at the recommended retail price, profiting from the trade discount (around 10 - 50%). In practice, if business were slow, you could end up chasing all over the country for very little profit.

The easiest way to get in to the spare parts trade is to specialise in used spares, combined with a small supply of fast moving (new) stock items. This trade can be lucrative; the dealer who offers £60 for a car which is about to be scrapped, will possibly sell the engine for around £70, and still have the rest of the car to sell!

Buying a range of cars necessitates the use of substantial premises, so the answer is to specialise. You can operate from your garage lock up, or even an outhouse. Stocking 'a bit of everything' will only frustrate inquirers who tend to ask for items not stocked, but with limited resources and space, you could still be able to carry virtually *all* spares for *one particular model*, ie. Metro, Fiesta, etc. As many parts are common to all variants of the model eg. 'L', Ghia, etc.), you will be capable of satisfying most customers without spreading your finances too thinly. If you specialise in rarer cars, you could advertise in the national press, and operate an additional postal service.

Non runners, M.O.T. failures, or 'bangers' can be found at auctions, or by advertising in the local press. Most cars are scrapped due to corrosion, rather than mechanical failure - ideal for you, as you wouldn't be selling major body components anyway. Once the car has been stripped, arrangements could be made with a scrap metal dealer for its uplift.

Guarantee your spares, you have nothing to lose. A ' no quibble' money back offer will attract hesitant customers, a refund given reluctantly will lose any repeat business. Some dishonest customers will fit the part that you supplied, then bring back their original faulty part, and try to claim a refund. The answer is to secretly mark any spares that you sell, for identification at a later date.

A different proven method of parts dealing, is to specialise in only one or two specific components (to suit all popular cars), ie. engines, gearboxes, diffs, computerised control boxes, etc. You could also offer a fitting service.

Sources for these components include answering private adverts (breaking for spares, garage clearouts, etc.), or even buying cheaply from scrapyards - a discount may be organised if you are a regular customer. If the dismantler is aware that you specialise, he will want to contact you as soon as a suitable vehicle arrives at his premises.

Again, your financial outlay will remain relatively modest, whilst being able to satisfy demand. A good guarantee (eg. three months) would mean that you could ask a higher price than your competitors. Although a gamble, the possibility of mechanical failure over this period is not very high.

New stock can be obtained from a motor trade cash and carry [42], and by using the new parts price guide published by QC Publications [43], you will be able to work out sensible prices for your used spares.

A second hand dealer's licence may be necessary before you start trading - check with your local council.

VANS

VAN HIRE WITH DRIVER
Luton vans prove popular with the owner driver. Ensure that you are covered for personal liability insurance.

Options for employment include: garden waste removal, contract deliveries with local companies, a 'panic' delivery service, and home delivery services from warehouses, supermarkets, and D.I.Y. stores.

Advertising costs may prove expensive to your new business; make sure your van is clearly signwritten (one of the cheapest ways to advertise), and display your business card in local shop windows.

VAN SALES ROUNDS
With most districts already covered, ice cream or grocery rounds will probably have to be bought as established businesses. Traders get very touchy about new competition on their 'patch'. Checks should be made with your local authority, and Environmental Health Department, as regards licensing regulations.

PARCEL DELIVERIES
This market is worth around £2 billion per year, and with increasing privatisation, is worth considering. Becoming part of an established network is one of the difficulties encountered; therefore an attractive option is that offered by a national franchise company. Amtrak, one such firm, has grown in the space of a few years to become one of the leading names in the industry. They operate by using trunker deliveries from a central sorting office, to depots throughout the country. It is from these that the franchisee delivers parcels within his own designated area. Full training will be given before starting business.

An initial franchise fee is payable (which could be financed through your bank), and you have to provide a diesel van of the type specified by the company. There are no premises or staff to pay for, and commission is paid monthly to you by Amtrak. As well as the benefits associated with working as a franchisee, your commission is guaranteed even in the event of non payment by the customer. For more information and an application form, contact Amtrak [49].

LIGHT REMOVALS

Luton bodied vans are most suited for light removals - the bigger the vehicle, the more scope there is for profit. If you intend using a bigger van, you may need to apply for a Goods Vehicle Operators Licence (see below). Those wishing to test the market (or can't afford the capital outlay) may wish to hire a self drive van for each job. This would be viable for bigger, long distance jobs, but less profitable for local work, unless rented at preferential rates.

Do your sums before quoting for any work. If you were successful, then demand for your services will increase, justifying the purchase of your own vehicle.

GOODS VEHICLE OPERATORS LICENCE

Should you intend to use a vehicle to carry goods, and the unladen weight is more than 1525kg. in the case of an unplated vehicle, or a gross plated weight of 3·5kg., you will need an operators licence.

The licence will be; 1) Restricted 2) Standard National, or 3) Standard International.

The Restricted licence allows you to carry your own goods in the U.K. or abroad, but not on behalf of other people.

The Standard National licence allows you to carry goods for other people (and yourself), while for journeys abroad, the International Standard licence applies.

Applications for these licences are made to the Traffic Area Office of your Department of Transport licensing authority (address in the 'phone book) covering the area in which you will be based. Apply first for the booklet GV 74, *A Guide for Operators*, which details the necessary application requirements.

You will apply on form GV 79 at least nine weeks before the licence is needed, and notice of your intent will have to be published in a local newspaper which covers the area of your proposed operating base.

A fee is payable, and when applying for the licence you must convince the authorities that you are 'fit' to hold the licence. Your past record will be taken in to account, so any criminal convictions have to be declared. You also need to show that you will have the use of suitable premises, and that you have the means practically, and financially, to keep your vehicles 'fit and serviceable'. You must be familiar with the rules regarding drivers' hours and vehicle overloading.

For Standard licence applicants, further requirements may apply, such as possession of a Certificate of Professional Competence (C.P.C.). Objections made by interested parties will be taken into account when the application is being considered.

A licence, if issued, may last up to five years, at the discretion of the licensing authority. If standards or conditions are not maintained, the authority may revoke, suspend, terminate, or limit the terms of any licence.

HIRING

Self drive car hire is theoretically quite straight forward when operated in conjunction with another existing forecourt business. However this market may be beyond the scope of the average man in the street; capital investment is particularly heavy, as you would need to work with a minimum of two or three new, or nearly new, cars. Insurance premiums are high. Written Terms and Conditions would have to be drawn up, entailing the use of a solicitor. Arrangements would have to be made with one of the major motoring organisations to cover mechanical failure, and recovery from anywhere in the country. Overheads include vehicle depreciation - higher with these newer cars. For the above reasons, it may be a safer option to contact one of the bigger national companies, with a view to becoming an agent.

Trailer hire is worth consideration. Capital investment is relatively low, there is less chance of mechanical breakdown, depreciation is low, although a pro forma legal contract for hires would have to be drawn up. Insurance should be relatively cheap, but you could make an additional charge for cover to the customer. Business premises may be humble, space even being rented from other business users, farms etc. Uplifting trailers that have broken down will not be practical if this has occurred far from home; a nationwide list of trailer repairers should be compiled. A large deposit will be required from the customer, in order that you see the trailer again!

Customer choice could range from the smallest box trailer to the most expensive van type enclosed trailer. Others include horseboxes, car transporters, 'A' frames, dollies, motorcycle transporters, and plant flatbed trailers. Caravans are a possibility, but there are a number of negative points - a) the hire season only lasts around three months in the year b) there is a high possibility of damage to upholstery, interior, or soft external alloy body panels c) these bigger trailers are more prone to accidental damage by inexperienced towing, and d) a high capital investment needed for caravans which are of an acceptable standard.

Stock could be bought new or second hand. If financially stretched, you could sub rent the more unusual trailers from other hirers. A sales and repair service could also be offered (see below).

TRAILERS

As suggested in the previous section, trailers may be considered a low budget and generally safe business investment. Relatively easy to sell, they are always in demand. Trailers are priced by condition rather than age - generally there are no second hand price guides, so opinions of value will differ between buyer and seller! Stock can be easily picked up from private seller via newspaper small ads, and sold on for a modest profit. On models over 6 or 7 years old, depreciation is virtually nil, so there is less pressure on you for a quick sale.

Your local motor trade cash and carry may offer new trailers in kit form. These are good products for resale; sometimes assembling component parts bought separately is not economically viable compared with the competition of cheaper retailers.

Repairs to trailers are relatively easy and straight forward. Wiring circuits are basic, and most component parts simply bolt on. Indespension [50] are the market leaders for parts, their address is at the back of this book.

OTHER IDEAS

The suggestions that follow may be suitable for full or part time work, or as a profitable hobby.

PROMOTIONAL & AGENCY WORK

Many car promotional and advertising agencies have sprung up, then disappeared without trace, over the last two decades. During research for this book, attempts to contact agencies from old advertisements 'drew a blank', which indicates the frailty of some of these schemes. In contrast, when managed properly, money can be made from advertising work, a good example being *Ad Trailers* - mobile billboards which are often seen in urban areas. One local agency operated by organising private cars to carry advertising material - fees being charged to the advertiser, and a commission paid to the driver.

Agencies are usually employed to supply video, film, and television studios with exotic, classic, and period vehicles. If you own a car that fits the above category, you may be able to earn cash by using the agency's services. Two leading, and long established companies are *Action Cars* [51], and *Action Vehicles* [52]. Operating throughout the U.K., and around the world, your car would be entered on their register. When a suitable job came up, your vehicle could then be collected, in some cases by trailer, and taken to the film location. For more details, contact the agencies.

SIGNWRITTEN ADVERTISING

If you have reservations about having your vehicle signwritten (which will entail a respray before selling or trading in the car), you might consider laser cut adhesive letters which are affixed to car windows, for easy removal at a later date. See under the heading 'Sign Writers' in Yellow Pages, for local suppliers.

An even cheaper method is to make large posters relevant to your trade, and place them in the side or rear car windows whenever it is left parked, and on public display.

MOTORING INFORMATION SERVICE - BT PREMIUM RATE LINES

If you wish to earn money by offering pre-recorded information, 'live' help lines, or give motoring advice to the public, you might consider setting up with a British Telecom 0891 Premium Rate Service. For each call (which is charged at a higher rate than the norm), BT will forward between 50 - 70% of the gross income. This service is only available to legitimate businesses, or information providers, and is overseen by a strict code of practice.

A connection charge is payable, as well as a quarterly rental (minimum 12 months). Fees are subject to V.A.T., and although reasonable, heavy expense is involved in advertising, and when buying special answering machinery. Obviously, if operating a 'live' line, you will need to employ staff. For details, ring the BT Premium Rate Service [53].

CAR MAGAZINES, BOOKS, AND PUBLICATIONS
A wealth of old car handbooks, workshop manuals, magazines, and other publications covering all forms of motor transport lie dormant in countless attics, boxes, and bookshelves. Charity shops, jumble, and car boot sales offer stock sources at budget prices. It may be of interest to note that old car sales brochures are fast becoming collector's items.

When sorted out and directed to the correct market, these books can be valuable. Once a reasonable collection has been assembled, advertise each in a related magazine, or better still, a club newsletter. For example, you may have some old Ford Prefect handbooks; these could be advertised for sale through a specialist magazine as published by the *Ford Prefect Fan Club*, or similar.

CAR BOOT SALES
Are obvious venues for the disposal of private garage clearouts, etc. Site fees are relatively cheap; increasingly, bona fide traders are accepted. Low overheads may help to encourage new business ventures.

FAST FOOD DELIVERIES
Ideal for students, or those looking for a little extra cash.

Find out if your local 'take away' restaurant has a home delivery service. If not, they would have nothing to lose and everything to gain by using your services. Approach them; their only outlay would be a hand written sign, displayed on the premises. You keep all delivery charges, while the shop still takes the full profit from the meals.

Using this system keeps the arrangement uncomplicated, and the restaurant gains custom from a wider sector of the market.

A light motorcycle would make your service more profitable, but in any case, check with your insurers that you have adequate cover.

MAGAZINE ADVERTISING PHOTO AGENT

This is the job that everyone wants! Working possibly on a self employed basis for a weekly car trader type paper. Supplied with a 35mm. camera and film, you will probably use your own car to visit those wishing to advertise their car in the magazine. Your job is to take the advert copy, photograph the car, and take the advertising fee!

Agents can get paid well, and the job looks easy, but can be hard work. Clients tend to wait until the advertising deadline before contacting the paper, thus creating a bottleneck of calls over two or three days. This is an interesting part time job requiring honesty and conscientiousness, coupled with the ability to handle cash.

Vacancies for photo agents are not often advertised, but you could contact your local magazine on the 'off chance'.

ORGANISING CHARITY TREASURE HUNTS

If you would like to raise money for a deserving charity, you could consider organising a car treasure hunt.

Each entrant will be charged a participation fee , with a non monetary prize for the winner. Place an advertisement in the local press detailing the entry fee, start location, and the cause being supported.

The treasure hunt should make a pleasant evening's entertainment for all concerned, and will hopefully boost the funds of your cause.

CARS - SKETCHES & PRINTS

For those with an artistic 'bent', painting or sketching vehicles from owner's photographs can be rewarding. Example advertisements of artists offering this service can be found in some of the collector's or classic car magazines.

To a competent artist, the job is relatively simple; copying is a lot easier than painting a real life model. An ideal size is A4, and pictures could be sold framed or unframed, priced accordingly. As an extension of this idea, some studios offer for sale standard prints of various models (eg. XJS, Metro, Sierra, etc.), to which the 'artist' will superimpose a registration number of the customer's choice.

G.R.P. BODY PANEL MOULDING

Glass reinforced plastic (glassfibre) moulding is a remarkably simple, if time consuming, process. Duplicates of original car or motorcycle body panels are relatively easy to manufacture, after experimentation and practice.

Books and leaflets offering technical advice, and step by step guides to laminating are available from Strand / Scott Bader [54], who are leading G.R.P. suppliers, having depots throughout the U.K.

CHILDREN'S PEDAL CARS

Collectors: In recent years there has been a keen interest in children's vintage pedal cars, resulting in inflated bid prices at major auction houses. When found, models are usually in a state of poor repair, or in unoriginal condition.

In contrast to other collector's pieces, pedal cars can realise good prices after careful restoration (which is usually straight forward). Libraries and museums may have to be consulted as regards original specifications such as colour, trim, etc.

The 'heyday' of children's pedal cars was between 1930 - 1960, when various clever models were produced - cars such as the Austin Junior Forty (J40), which included a detailed dummy engine; and the Austin Pathfinder Special which sported pneumatic tyres, working handbrake, removable bonnet, which also revealed a dummy engine, fitted with real spark plugs!

A recommended book on the subject is the cheap and informative Shire Album *'Children's Cars' (number 178)*, which gives a detailed illustrated subject history, from the turn of the century to date. Places to visit, and further recommended reading is listed. Contact Shire Books [55] for further information, and lists of other obscure and interesting titles available, many of interest to the motoring enthusiast.

Construction: Designing and building battery cars could be profitable, good fun, and relatively simple after sources for component parts were identified. Electrically propelled children's cars offered for retail sale in toy shops can be surprisingly expensive.

For first attempts, you might wish to work from commercially available plans, such as those offered by Real Life Toys of Sheffield [56], who will supply kits, assembled models, and parts.

APPENDIX

ALL MOTORING SERVICES

MONEY SAVERS

GENERAL TIPS
* Everyone knows that unleaded fuel is good for the environment, and saves money, but some motorists are unaware that the car's conversion usually involves no more than retarding the ignition timing 2 or 3 degrees - a job which takes about 10 minutes. Many garages will charge a minimum of one hours labour for this adjustment.

Before making any alteration to your car's specification, check with your main dealer / workshop manual - some vehicles are not suited to run unleaded, or other adjustments may be required.

* On the subject of petrol savers, 'WHICH', the magazine published by the Consumers' Association [4], has tested various bolt on fuel economisers. The results were distinctly discouraging, and the conclusion was that the best product was the magazine's own accessory; the cost was nil, and it consisted of a list of tips such as returning the choke as soon as possible after starting the engine, reading the road ahead to avoid heavy braking, using the highest suitable gear, accelerating gently, and ensuring correct tyre pressures etc., etc.

* Using a roofrack is very uneconomical; the drag factor drastically increases petrol consumption.

* Cheap engine oil is a false economy, it causes sludging, carbon deposits, and premature engine wear. Not recommended!

* Reference was made earlier to the price of exhaust systems. Provided you can remove the old one, fitment of the new is usually very easy - check the take away price at a motor factor against that of one fitted by an exhaust centre.

SAVE ON MOTOR INSURANCE
* Advanced driving certificates, and car alarms / immobilisers fitted, will reduce the premiums charged by certain companies.

* Check if your insurer offers 'classic status' cover for older vehicles. Big savings can be made with these policies, but you might be bound by limitation clauses regarding allowable mileage, etc.

* Some owners continue with comprehensive insurance on a car that has depreciated in value - almost down to the same level as that of the annual premium! If cover was then changed to third party only, then, provided there were no claims for at least a year, these drivers would be in profit - even in the event of a total loss of the car.

* Don't ever lie when filling in the insurance proposal form. Although this may save a few pounds in the short term, you may find that in the event of a claim, the policy is void.

* Remember, if selling your car (and not immediately replacing it), to cancel or suspend the policy. You should be entitled to a refund for the remaining period. Alternatively, the insurance will be 'frozen' until transferred to the replacement car.

* Loyalty does not pay! It is amazing the amount of motorists who blindly re-insure their car with the same company year after year. Get plenty of insurance quotes at renewal time; the difference in premium prices can be surprising. Make full use of brokers.

* **After shopping around** for the best quote, it could pay you to contact Motoradvice [57], an enterprising company that will probably undercut your quoted figure. After paying the premium, Motoradvice will (in most cases) then apply a service charge, being *25% of the difference saved.* What have you to lose?

LIFT POOLS
Until relatively recently, car insurance could be invalidated if a driver accepted money from colleagues (eg. sharing a lift to work, etc.), to go towards petrol expenses, etc. They may now contribute, provided the lift is for 'a social purpose', and no profit is made.

LOSS OF LICENCE - INSURANCE
In the event of a licence being lost through medical grounds, disqualification, or in the event of having to hire a car (until reimbursed by your insurers), because your own was stolen, vandalised, or accident damaged - a suitable private insurance plan may be the answer. Note that some companies will not provide cover for drink / drug offences, or those directly associated with dangerous, reckless, or careless driving.

A scheme comparable with the above is offered with the St. Christopher Driving Plan [58], address at the back of the book.

MECHANICAL BREAKDOWN RECOVERY INSURANCE
Annual subscriptions to the major motoring (road side assistance) organisations have risen steadily over the years. Peace of mind is expensive for a recovery service that you may never need to use!

An alternative to membership exists in breakdown recovery insurance. Boncaster Ltd [59], offer their AUTOAID scheme at a *fraction of the fee charged by the main two motoring associations.* The 1994 premium was £15 for owner and spouse!

Roadside repairs, or recovery costs are paid in the first instance by you, then the bills are forwarded to Boncaster for reimbursement - should you pay for the recovery with a credit card, you might never be out of pocket!

Autoaid is effected with Lloyds underwriters, and covers:-

a) *Roadside repairs*

b) *Recovery of vehicle and up to five passengers to a chosen destination in mainland England, Scotland, or Wales*

c) *Alternative transport*

d) *Overnight accommodation*

e) *Breakdowns at home*

f) *Any vehicle (less than 20 years old)*

g) *Trailers or caravans attached at the time of breakdown*

h) *Reimbursement of charges from special areas such as toll tunnels and bridges*

SECURING TRADE DISCOUNTS

Never pay full price for anything unless you are forced to! Whether you are dealing with the motor, building, or any other trade, discounted prices are nearly always available upon production of suitable evidence of trading. This may take the form of an official order, or just a business card, both of which are available easily and cheaply - well worth having for years of discounted goods and services.

Should you be on a tight budget, or require only a small quantity of cards, try using the instant business card machines located in shopping malls and motorway service stations. If official order forms are necessary, organise a rubber stamp and pad for use with a duplicate order book, available from stationers. When only a small quantity of headed paper is needed, these can easily be produced by means of 'Letraset' type dry transfers to produce a mock up, which is then photocopied to produce the finished product - modern copying machines offer near perfect results.

ADMINISTERING YOUR BUSINESS

Red tape and statutory regulations all make the running of any business that little bit harder. Dealing with income tax, National Insurance, and V.A.T. can be complicated, every trader having his own requirements. With this in mind, the following information must be treated as a very rough guide (yet again!); further professional advice should be sought to suit your own individual needs. Another theme which has run throughout this book: *always* make sure that you have adequate insurance cover for any business undertaking, particularly public and employee's liability, where appropriate. Any insurance broker will explain the extent of cover required.

Working as 'the boss', you will probably have more job satisfaction, pay lower National Insurance contributions, receive tax benefits, and be better off financially. In contrast, hours may be long, you will receive no holiday pay, no redundancy payments, and no unemployment benefit in the event of your business failing.

SOLE TRADER
Anyone can set up as a sole trader. There is no legal distinction between you and your business. Liability rests upon you personally, and creditors can claim against your own possessions or home, if you get in to financial difficulties.

As a sole trader. you may employ staff, and enjoy the benefits of any profit. Your name should appear on business stationary and invoices. There is no requirement to register the business with Companies House [60], and although there are no rules on the records you keep, you should retain all receipts, and details of sales, for tax purposes.

PARTNERSHIP
Consisting of two or more sole traders (but not exceeding 20), similar rules to the above apply. Before entering a partnership, you should draw up a legal contract - many lifelong friends have been lost this way! Tax is payable on the collective, rather than the individual profit. Advantages are the pooling of resources and skills, the disadvantage being that everyone suffers if one individual makes a serious judgement error.

LIMITED LIABILITY COMPANY

The 'shareholders' own the company, but are not liable for its debts. Any liability is limited to the share capital contributed. Income tax is only payable on salaries drawn, and accounts are submitted to Companies House, where they are available for public scrutiny. *The rules for trading are involved, so contact your accountant for further advice.*

CO-OPERATIVE

Most co-operatives are set up as limited liability companies. You would own, control and work jointly with others, decisions and responsibilities are shared. For further information, contact I.C.O.M. - the Industrial Common Ownership Movement [61].

ENTERPRISE TRUSTS - SMALL FIRMS SERVICE

Training and Enterprise Councils, or Local Enterprise Companies (L.E.C.s) offer various free booklets (including an accounting guide) to those intending to start and run their own new business. Information, advice, and possibly free counselling sessions may also be available. To obtain further information on what is available near you, contact your local Jobcentre, who will refer you to the appropriate body.

NEW BUSINESS STARTUP SCHEMES

The New Business (or Enterprise Allowance) Scheme is payable for up to 52 weeks to those wishing to start up their own business, and is in addition to any profits made.

Applicants should be in receipt of Unemployment Benefit, or Income Support. They should be aged 18 - 65 and able to invest £1,000 in the new venture (a bank loan may be acceptable). The business must be approved by the Department of Employment, and the applicant must agree to work full time (at least 36 hours per week).

Trading must not have commenced before entering the scheme, and an 'awareness day', organised by the Department will have to be attended before qualifying as a beneficiary.

The scheme also provides access to free business advice and training, to help overcome the problems encountered in the every day running of the new business. Free counselling sessions may be available from a local Enterprise Trust (see above). To find out more, contact your local Jobcentre, or Department of Employment.

THE ACCOUNTANT

Ignoring the services of an accountant (at least in the first year of trading) can prove to be a false economy. Although not cheap, he may save you more than his fee in respect of any tax that you might unwittingly pay the Inland Revenue. The accountant will come to an agreement with the authorities, and claim any relief that you are due. An accountant can also be an excellent source for business advice.

Bear in mind that whether or not the accountant has audited your books, **it is still you who is ultimately responsible** for their accuracy, and the declaration of any profits.

When choosing your accountant, biggest does not necessarily mean best. Get an estimate of what you might be charged - some accountants can be very expensive indeed. Ask around for recommendations from friends or business acquaintances.

THE BANK MANAGER

Most customers are frightened off before they even see the bank manager-remember that his business is to help *you* make money, which will ultimately be to his benefit as well. There is usually no charge for private consultations, but if in connection with business affairs, there may be a nominal fee.

The bank manager should be able to advise on straight forward tax matters; anything more complicated may necessitate an appointment with his tax specialist.

If looking for a loan to set up trading, you will have to present a convincing case - you may need to organise a business plan for his perusal. It is pointless to approach the manager for loan facilities, unless a genuinely viable proposition is put forward.

For help preparing the plan, contact the small firms service at your local Enterprise Trust or Company (see page 140).

The bank will, in addition, require some form of security to cover the loan (refer to pages 10-11). Some examples of security might be:-

a) *Your house*

b) *A life insurance policy (find out the surrender value), but the bank may insist on a policy being in force as a condition of the loan to cover itself in the event of your premature death*

c) *Guarantees from friends, relatives, or partners*

d) *Stocks or shares - the bank may wish you to put up a higher value than that of the loan, due to stock market fluctuations*

BANK CHARGES

These charges may be quite substantial for the small trader, and are often overlooked until the arrival of the bank statement!

If few cheques are drawn during each calendar month, it may be worthwhile consulting a building society (such as the Nationwide), that offers a business cheque account. Provided the stipulated amount of withdrawals is not exceeded within each month, not only will the banking be free of charge, but in addition, interest will be *credited* to the account!

CREDIT CARD PAYMENT FACILITIES

Acceptance of 'plastic money' eliminates the risks traditionally associated with cheque payments for goods and services. Worth £40 billion of sales each year, credit cards can tempt the holder to spend more - usually on impulse buys.

There is a service charge to the retailer on each transaction, in addition to an initial set up fee. For further details, telephone Barclays Merchant Bank Services [62], or contact the local branch of your bank.

INCOME TAX

As soon as you start working for yourself, you should inform your local tax office (address in the 'phone book, under Inland Revenue). It is possible that it is not the same office that handled your Pay As You Earn (P.A.Y.E). contributions. Ask for their helpful guide book IR 28, 'Starting In Business'.

Tax will be paid under 'Schedule D', which offers; a) a wide range of business expenses to set against the tax payable, b) a considerable time delay between earning the money and paying the due figure, ie. the tax is paid in arrears, and c) payment is made by two instalments, in January and July.

When operating your business from home, a proportion of your domestic expenses may be set against tax.

Records will have to be kept, eg. purchase and sales ledgers, also all receipts for legitimate expenses should be retained as proof.

'Schedule D' may still apply (benefiting from its advantages) via a separate system, if you intend trading part time while still working for an employer and remaining on 'Schedule E' (P.A.Y.E).

The help of an accountant at an early stage is advisable.

NATIONAL INSURANCE

In order that you may receive state benefits, most people pay National Insurance contributions. If intending to go self employed, contact your local Department of Social Security office (address in the 'phone book). There are free information leaflets available, and you will be advised which category of contributions payment class applies to you.

Class 1 will entitle you to sickness, old age, and unemployment benefits.

Class 2 (Self employed) entitles you to sickness, and old age but *not* unemployment benefit.

Class 3 (Not employed, but not claiming unemployment benefit) - entitlement to old age benefit only.

Class 4 may have to be paid when you are self employed, and earnings are above a specified limit.

When paying Class 2, most contributors elect to pay by direct debit, thus simplifying the transactions. Although payments are flat rate, they fluctuate slightly each month. At the beginning of each tax year, contributors receive notice of each monthly rate, and the exact dates of deduction.

VALUE ADDED TAX - GENERAL INFORMATION

You are required to be V.A.T. registered when your annual business turnover exceeds (or looks as though it might exceed) the stipulated figure set by Customs and Excise. This being the case, you will be obliged to charge the appropriate amount of Value Added Tax on all or some of your goods and services. Set against this is the V.A.T. charged to you by your suppliers.

Regulations are complex, so it is important that you contact your local V.A.T. office (see under Customs & Excise in the 'phone book). Ask them for a free copy of their book *'Notice 700 - The V.A.T. Guide'.* Further advice may be sought from your accountant as soon as possible; ignorance is no excuse in law, and the penalties for flouting V.A.T. rules can be both severe and immediate.

A special tax scheme operates for used car traders, refer to page 56 *et seq.*

PLANNING PERMISSION

Throughout this book, references have been made to the fact that the appropriate planning permission is obtained from your local authority, before you start trading. Local councils (or building societies, if the property is mortgaged) may prohibit or impose conditions on the running of a business from certain properties. Should this be the case, you could consider using a 'convenience' office address. This service gives you the benefit of a different postal address, and will probably offer a telephone answering service. See Yellow Pages - 'Accommodation Agents - Business', for a service near you.

LOW RENT OPTIONS

Low rent or even free limited use of business premises is sometimes offered by local district authorities in financially depressed areas. These advantages are directed to new or existing companies that have the potential to employ a local workforce, and are considering a move into the area. Contact your local business development agency, or local council for possible options.

AND FINALLY...

Address references in the book have been marked ' [ ] ', and the number within the brackets corresponds to an address or telephone number that follows in the next section.

I hope you have enjoyed the book and found it informative. To those of you who wish to put some ideas into practice, it only remains for me to wish you Good Luck!

[1] Ford Motor Co. Ltd.,
Tax Free Sales,
Wintersells Road,
Byfleet,
Surrey KT14 7LF.
Tel: 01932 335113

[2] Nationwide Used Car
Arbitration Scheme (NUCA)
Freephone: 0800 834741

[3] HP Information PLC
Private Purchasers Help Line
Tel: 01722 422422

[4] The Consumer's Association,
PO Box 44,
Hertford X, SG14 1SH.
Freephone: 0800 252100

[5] RAC Motoring Services,
Spectrum,
Bond Street,
Bristol BS99 1RB.
Tel: 01179 232444

RAC Vehicle Inspections
Freephone: 0800 333660

[6] The Automobile Association,
Fanum House,
Basingstoke,
Hampshire RG21 2EA.
Tel: 0345 500600

AA Vehicle Inspections
Tel: 0345 500610

[7] HMSO,
PO Box 276,
London SWH 5DT.
Tel: 0171 873 9090

[8] Standard Forms Ltd.,
Unit 10,
Romsey Industrial Estate,
Greatbridge Road,
Romsey,
Hampshire SO51 OHR.
Tel: 01794 517206

[9] Glass's Guide Service Ltd.,
Elgin House,
St. Georges Avenue,
Weybridge, KT13 OBX.
Tel: 01932 853211

[10] CAP Nationwide Motor
Research Ltd.,
CAP House,
Carleton Road,
Skipton, BD23 2BE.
Tel: 01756 700666

[11] Motorists Guide,
Foxpride Ltd.,
67, Tyrell Street,
Leicester LE3 5SB.
Tel: 01162 511393

[12] RAC Mechanical Insurance
Services Ltd.,
PO Box 151,
Tunbridge Wells,
Kent TN1 1RQ.
Tel: 01892 510606

[12] Car Care Plan Ltd.,
Bramley District Centre,
Leeds LS13 2EJ.
Tel: 01132 562133

[13] Wagon Finance Ltd.,
302, Ringing Low Road,
Sheffield S11 7PX.
Tel: 01142 303400

[13] Lloyds Bowmaker Ltd.,
Finance House,
Orchard Brae,
Edinburgh EH4 1PF.
Tel: 0131 332 2451

[14] SMTA Ltd.,
3, Palmerston Place,
Edinburgh EH12 5AF.
Tel: 0131 225 3643

[15] National Conciliation Service,
RMIF,
9, North Street,
Rugby CV21 2AB.
Tel: 01788 576465

Other enquiries:
RMIF,
201, Great Portland Street,
London W1N 6AB.
Tel: 0171 580 9122

[16] SMMT,
Forbes House,
Halkin Street,
London SW1X 7DS.
Tel: 0171 235 7000

[17] Bollington Insurance,
51, Palmerston Street,
Bollington,
Macclesfield,
Cheshire SK10 5PW.
Tel: 01625 574342

[17] Clegg, Gifford & Co.,
199, Ilford Lane,
Ilford IG1 2RX.
Tel: 0181 478 6821

[17] TS Uttley & Son,
13, Bridge Street,
Southport,
Lancs. PR8 1BW.
Tel: 01704 538524

[18] VMC Ltd.,
PO Box 16,
Marple,
Stockport,
Cheshire SK6 7HD.
Tel: 01663 766047

[19] HP Information Plc,
Dolphin House,
PO Box 61,
New Street,
Salisbury,
Wilts. SP1 2TB.
Tel: 01722 413434
Sales Info: 01722 412888

Private Buyer Enquiries:
01722 422422

[20] Portfolio Marketing,
Elland Lane,
Elland,
W.Yorkshire HX5 9DU.
Tel: 01422 310044

[21] *For 'V' Bar:*
Midas,
Engine Lane,
Horbury Bridge,
Wakefield WF4 5NH.
Tel: 01924 266266

[21] *Towing Dollies/ 'A' Frames:*
McArdle Fabrications,
Unit 3,
Shilton Industrial Estate,
Coventry CV9 9QL.
Tel: 01203 612463

[22] 'AUTO TRADER' Magazines
Regions Covered:
Scottish
North West
North East
Yorkshire
Midland
Anglia
Thames Valley
North London
South London
Southern
South West
Western
(See page 152)

[23] Popplewells Alignment Ltd.,
High Road,
Thornwood,
Epping,
Essex CM16 6LP.
Tel: 01992 561571

[24] Autolign Ltd.,
JBJ Business Park,
Northampton Road,
Blisworth,
Northampton NN7 3DW.
Tel: 01604 859424

[25] Society of Motor Auctions,
Sunrise House,
Hulley Road,
Hurdsfield Ind. Estate,
Macclesfield,
Cheshire SK10 2LP.
Tel: 01625 502997

[26] ADT Auctions,
Expedier House,
Portsmouth Road,
Hindhead,
Surrey GU26 6TJ.
Tel: 01428 607440
(Sales Nationally)

[26] Central Motor Auctions,
Central House,
Pontefract House,
Rothwell,
Leeds LS26 OJE.
Tel: 01132 820707
(Sales Nationally)

[26] National Car Auctions,
Frating,
Colchester,
Essex CO7 8TD.
Tel: 01206 250230
(Sales Nationally)

[27] DVLA (DVLC),
Swansea SA99 1AL.
Driver Licence Enq.
Tel: 01792 772151

Vehicle Related Enq.
Tel: 01792 772134

[28] British Franchise Assn.,
Thames View,
Newtown Road,
Henley on Thames,
Oxon. RG9 1HG.
Tel: 01491 578050

[29] BSM,
81 / 87, Hartfield Road,
London SW19 3TJ.
Tel: 0181 540 8262

[30] DIA,
Safety House,
Beddington Farm Rd.,
Croydon CR0 4XZ.
Tel: 0181 665 5151

[31] MSA,
182a, Heaton Moor Rd.,
Stockport,
Cheshire SK4 4DU.
Tel: 0161 443 1611

[32] RCM Marketing Ltd.,
80, Tenter Road,
Moulton Park Business
Centre,
Northampton NN3 6AX.
Tel: 01604 790890

[33] He-Man Equipment Ltd.,
Princes Street,
Northam,
Southampton SO14 5RP.
Tel: 01703 226952

[34] DSA,
Stanley House,
Talbot Street,
Nottingham NG1 5GU.
Tel: 01159 474222

[35] DVLA Registration Marks
Sales 'Hotline'
Tel: 0181 200 6565

[36] CNDA,
201, Great Portland Street,
London W1N 6AB.
Tel: 0171 580 9122

[37] Graham Gomm (Ins.Brokers),
260,Wellington Road,
Perry Barr,
Birmingham B20 6PU.
Tel: 0121 356 5310

[38] Public Carriage Office,
15, Penton Street,
London N1 9PU.
Tel: 0171 230 1633

[39] National Federation of
Taxi Cab Associations,
The Views,
George Street,
Huntingdon,
Cambs. PE8 6BR.
Tel: 01480 432223

[40] Autoglym,
Works Road,
Letchworth,
Herts. SG6 1LU.
Tel: 01462 677766

[41] Autosheen,
21 / 25, Sanders Road,
Finedon Rd. Estate,
Wellingborough,
Northants. NN8 4NL.
Tel: 01933 272347

[42] Maccess Ltd.,
Spen Lane,
Cleckheaton,
W.Yorkshire BD19 4PG.
Tel: 01274 870241
(Branches Nationwide)

[43] QC Publications,
Kersey Hall,
Combs,
Stowmarket,
Suffolk IP14 2EZ.
Tel: 01449 676620

[44] Holts Spraymatch Centres
Enquiry Line (Bodycare)
Tel: 01925 633803

[45] CP Witter Ltd.,
18, Canalside,
Chester CH1 3LL.
Tel: 01244 341166

[46] Glass Aid (UK),
Unit 4,
Ambassador Ind. Estate,
Airfield Road,
Christchurch,
Dorset BH23 3TG.
Tel: 01202 499619

[47] Hometune,
Carlton House,
Nantwich Road,
Calveley,
Cheshire CW6 9JW.
Tel: 01829 260030

[48] Computa Tune,
9, Petre Road,
Clayton Park,
Clayton-le-Moors,
Accrington,
Lancs. BB5 5JB.
Tel: 01254 391792
(Scotland: 01698 263449)

[49] Amtrak Express Parcels Ltd.,
Company House,
Tower Hill,
Bristol BS2 OEQ.
Tel: 01179 272002

[50] Indespension,
Belmont Road,
Bolton,
Lancs. BL1 7AQ.
Tel: 01204 309797

[51] Action Cars,
Units 3 + 4,
Rosslyn Crescent,
Harrow,
HA1 2SP.
Tel: 0181 863 6889

[52] Action Vehicles,
Shepperton Film Studios,
Shepperton,
Middlesex TW17 0QD.
Tel: 01932 566766

[53] BT Premium Rate Information
Freephone: 0800 333300

[54] Scott Bader Co. Ltd.,
Wollaston Hall,
Wollaston,
Wellingborough,
Northants.NN29 7RL.
Tel: 01933 664455
(Branches Nationwide)

[55] Shire Publications Ltd.,
Cromwell House,
Church Street,
Princes Risborough,
Aylesbury,
Bucks. HP17 9AJ.
Tel: 01844 344301

[56] Real Life Toys,
New Street,
Holbrook Ind. Estate,
Halfway,
Sheffield S19 5GH.
Tel: 01142 510300

[57] Motoradvice,
8, Queens Road,
London SE15 2PT.
Tel: 0171 639 9734

[58] St.Christopher Driver Plan,
PO Box 179,
Douglas,
Isle of Man.
Tel: 01624 629494

[59] Boncaster Ltd.,
Library House,
New Road,
Brentwood,
Essex CM14 4GD.
Tel: 01277 200670

[60] Companies House,
Crown Way,
Cardiff CF4 3UZ.
Tel: 01222 380801

[61] ICOM,
Vassalli House,
20, Central Road,
Leeds LS1 6DE.
Tel: 01132 461738

[62] Credit Card Acceptance:
Contact your own bank, or
Barclays,
Freephone: 0800 616161

REGIONAL AUTO TRADER MAGAZINES

SCOTTISH AUTO TRADER, 14, Dalzell Drive, Motherwell. Tel: 01698 258811.

NORTH WEST AUTO TRADER, Unit 1, Bewsey Industrial Estate, Catherine Street, Bewsey, Warrington, Lancs. Tel: 01925 33000.

NORTH EAST AUTO TRADER, Auto Trader House, High Street, Gateshead. Tel: 0191 490 0999.

YORKSHIRE AUTO TRADER, Munro House, Duke Street, Leeds. Tel: 01132 430300.

MIDLAND AUTO TRADER, Auto Trader House, 14 -16, Phoenix Park, Avenue Road, Nechells, Birmingham. Tel: 0121 333 3311 / 3111 / 3380.

ANGLIA AUTO TRADER, 84, St. Benedicts Street, Ipswich. Tel: 01473 287555.

THAMES VALLEY AUTO TRADER, Auto Trader House, Danehill, Cutbush Park, Lower Early, Reading. Tel: 01734 312222 / 313131.

NORTH LONDON AUTO TRADER, 43-45, High Road, Bushey Heath, Herts. Tel: 0181 950 9900.

SOUTH LONDON AUTO TRADER, Auto Trader House, 2, Jubilee Way, Merton, London SW19 3XD. Tel: 0181 543 8000.

SOUTHERN AUTO TRADER, 421-427, Millbrook Road, Millbrook, Southampton. Tel: 01703 701033.

SOUTH WEST AUTO TRADER, Auto Trader House, Babbage Road, Totnes Industrial Estate, Totnes, Devon. Tel: 01803 862248.

WESTERN AUTO TRADER, 1, Buckingham Court, Beaufort Business Park, Bradley Stoke North, Bristol. Tel: 01454 616161.

SALES OFFICES:

AUTO TRADER NATIONAL SALES LTD., Unit 5, 50, Windsor Avenue, London SW19. Tel: 0181 543 2323.

AUTO TRADER NATIONAL SALES (NORTH), Suite 2, Mallard Court, Broadway, Salford. Tel: 0161 877 9977.

NOTES / ADDENDUM